Lecture Notes in Computer Sci‹

Edited by G. Goos, J. Hartmanis and J. van

Springer

Berlin
Heidelberg
New York
Barcelona
Hong Kong
London
Milan
Paris
Singapore
Tokyo

Anthony Ambler   Seraphin B. Calo
Gautam Kar (Eds.)

Services Management in Intelligent Networks

11th IFIP/IEEE International Workshop
on Distributed Systems:
Operations and Management, DSOM 2000
Austin, TX, USA, December 4-6, 2000
Proceedings

 Springer

Series Editors

Gerhard Goos, Karlsruhe University, Germany
Juris Hartmanis, Cornell University, NY, USA
Jan van Leeuwen, Utrecht University, The Netherlands

Volume Editors

Anthony Ambler
The University of Texas
Department of Electrical and Computer Engineering
Austin, TX 78712, USA
E-mail: ambler@ece.utexas.edu

Seraphin B. Calo
Gautam Kar
T.J. Watson Research Center, IBM Research Division
30 Saw Mill River Road, Hawthorne, NY 10532, USA
E-mail: {scalo/gkar}@us.ibm.com

Cataloging-in-Publication Data applied for

Die Deutsche Bibliothek - CIP-Einheitsaufnahme

Services management in intelligent networks : proceedings / 11th
IFIP/IEEE International Workshop on Distributed Systems: Operations
and Management, DSOM 2000, Austin, TX, USA, December 4 - 6, 2000.
Anthony Ambler ... (ed.). - Berlin ; Heidelberg ; New York ; Barcelona ;
Hong Kong ; London ; Milan ; Paris ; Singapore ; Tokyo : Springer, 2000
(Lecture notes in computer science ; Vol. 1960)
  ISBN 3-540-41427-4

CR Subject Classification (1998): C.2, K.6, D.1.3, D.4.4

ISSN 0302-9743
ISBN 3-540-41427-4 Springer-Verlag Berlin Heidelberg New York

Springer-Verlag Berlin Heidelberg New York
a member of BertelsmannSpringer Science+Business Media GmbH
© Springer-Verlag Berlin Heidelberg 2000
Printed in Germany

Typesetting: Camera-ready by author, data conversion by Christian Grosche, Hamburg
Printed on acid-free paper   SPIN: 10781030   06/3142   5 4 3 2 1 0

Preface

This volume of the Lecture Notes in Computer Science series contains all the papers accepted for presentation at the 11th IFIP/IEEE International Workshop on Distributed Systems: Operations & Management (DSOM 2000), which was held at the Thompson Conference Center, The University of Texas, Austin, Texas, USA, from 4 to 6 December, 2000.

DSOM 2000 is the eleventh workshop in a series of annual workshops and it follows in the footsteps of highly successful previous meetings, the most recent of which were held in Zurich, Switzerland (DSOM'99), Delaware, USA (DSOM'98) and Sydney, Australia (DSOM'97). The goal of the DSOM workshops is to bring together researchers in the areas of network, systems and services management, from both industry and academia, to discuss recent advances and foster future growth in this field. In contrast to the larger management symposia, such as IM (Integrated Management) and NOMS (Network Operations and Management Symposium), the DSOM workshops are organized as single-track programs in order to stimulate interaction among participants.

The specific focus of DSOM 2000 is "Services Management in Intelligent Networks", reflecting the current interest and development in the field of distributed, networked application services, their definition, operation and management. Most of the papers presented at the workshop address some important aspect of this broad problem area. This year we have departed slightly from the traditional DSOM format. We have introduced two sessions devoted to the presentation and discussion of papers that describe work-in-progress.

The level of interest in DSOM has been growing steadily over the years. This year we were fortunate to receive 65 high quality papers from 14 countries, of which 21 were selected for the seven technical sessions. In addition, we received 19 work-in-progress papers, of which 12 were selected.

This workshop owes its success to all the members of the technical program committee, who did an excellent job of encouraging their colleagues in the field to submit high quality papers, and who devoted a lot of their time to help create an outstanding technical program. We thank them sincerely. We are also very grateful to the volunteer reviewers who gave generously of their time to make the review process effective. The DSOM 2000 website and the automated paper submission and review system were developed and hosted at POSTECH, Korea, under the supervision of James Hong and Sook-Hyun Ryu. We are thankful to them for providing this facility. And finally, we would like to express our thanks to Alexander Keller for helping us edit and put together this volume.

October 2000

Anthony Ambler
Seraphin B. Calo
Gautam Kar

Organization

General Chair

Seraphin B. Calo   *IBM T.J. Watson Research Center, U.S.A.*

Technical Program Co-chairs

Anthony Ambler   *University of Texas at Austin, U.S.A.*
Gautam Kar   *IBM T.J. Watson Research Center, U.S.A.*

Liaisons

IFIP TC WG6.6: Wolfgang Zimmer, GMD First, Germany
IEEE ComSoc: Douglas Zuckerman, Telcordia Technologies, U.S.A.

DSOM 2000 Supporters and Patrons

IBM Research Division, Yorktown Heights, New York, U.S.A.

IEEE, Institute of Electrical and Electronics Engineers

IFIP, International Federation for Information Processing

Tivoli Systems Inc., Austin, Texas, U.S.A.

Program Committee

Sebastian Abeck	*University of Karlsruhe, Germany*
Nikos Anerousis	*Voicemate, U.S.A.*
Enrico Bagnasco	*CSELT, Italy*
Suzanne Barber	*University of Texas, U.S.A.*
Raouf Boutaba	*University of Waterloo, Canada*
Marcus Brunner	*NEC Europe, Germany*
Rob Davison	*British Telecom, U.K.*
Metin Feridun	*IBM Research, Switzerland*
Kurt Geihs	*University of Frankfurt, Germany*
German S. Goldszmidt	*IBM T.J. Watson Research Center, U.S.A.*
Sigmund Handelman	*IBM T.J. Watson Research Center, U.S.A.*
Heinz-Gerd Hegering	*University of Munich, Germany*
Joseph Hellerstein	*IBM T.J. Watson Research Center, U.S.A.*
James Hong	*POSTECH, Korea*
Gabriel Jakobson	*GTE Laboratories, U.S.A.*
Irene Katzela	*Lucent Netcare, Canada*
Ryutaro Kawamura	*NTT, Japan*
Alexander Keller	*IBM T.J. Watson Research Center, U.S.A.*
Yoshiaki Kiriha	*NEC, Japan*
Emil Lupu	*Imperial College, U.K.*
Hanan Lutfiyya	*University of Western Ontario, Canada*
Kenneth Lutz	*Telcordia Technologies, U.S.A.*
Thomas Magedanz	*IKV++ GmbH, Germany*
Jean-Philippe Martin-Flatin	*EPFL, Switzerland*
Subrata Mazumdar	*Avaya Communication, U.S.A.*
Branislav Meandzija	*General Instrument Corporation, U.S.A.*
George Pavlou	*University of Surrey, U.K.*
Aiko Pras	*University of Twente, The Netherlands*
Pradeep Ray	*University of New South Wales, Australia*
Jürgen Schönwälder	*TU Braunschweig, Germany*
Adarshpal Sethi	*University of Delaware, U.S.A.*
Morris Sloman	*Imperial College, U.K.*
Rolf Stadler	*Columbia University, U.S.A.*
Burkhard Stiller	*ETH Zürich, Switzerland*
Carlos B. Westphall	*Federal University of Santa Catarina, Brazil*
Wolfgang Zimmer	*GMD FIRST, Germany*
Simon Znaty	*EFORT, France*
Douglas Zuckerman	*Telcordia Technologies, U.S.A.*

Reviewers

The task of reviewing the papers submitted to DSOM 2000 was extremely important. It is therefore a great pleasure to thank all the reviewers listed below for their serious and detailed comments. The list includes both members of the DSOM Committees and volunteer reviewers. Their efforts were key in maintaining the high quality of the workshop.

Sebastian Abeck
Anthony Ambler
Nikos Anerousis
Karen Appleby
Juan-Ignacio Asensio
Enrico Bagnasco
Suzanne Barber
Mark Bearden
Raouf Boutaba
David Breitgand
Marcus Brunner
Seraphin B. Calo
Ho-Yen Chang
Rob Davison
Gabi Dreo Rodosek
Toru Egashira
Metin Feridun
Luciano Gaspary
Kurt Geihs
German S. Goldszmidt
Lisandro Granville
Abdelhakim Hafid
Sigmund Handelman
Heinz-Gerd Hegering
Joseph Hellerstein
Christian Hoertnagl
James Won-Ki Hong
Gabriel Jakobson
Gautam Kar
George Karetsos
Olaf Kath
Ryutaro Kawamura
Alexander Keller
Jae-Young Kim
Andreas Kind
Yoshiaki Kiriha
Armin Kittel

Yao Liang
Jorge E. Lopez-de-Vergara
Emil Lupu
Hanan Lutfiyya
Kenneth Lutz
Sheng Ma
Thomas Magedanz
Ian Marshall
Jean-Philippe Martin-Flatin
Subrata Mazumdar
Branislav Meandzija
Marcos Novaes
Yongseok Park
George Pavlou
Aiko Pras
Xuesong Qiu
Pradeep Ray
Jürgen Schönwälder
Adarshpal Sethi
Chien-Chung Shen
Ana Cristina Silva
Edson Silva Jr.
Morris Sloman
Ron A. M. Sprenkels
Rolf Stadler
Burkhard Stiller
Daniel Swinehart
Jorge Tellez
Szymon Trocha
Venu Vasudevan
Elvis Vieira
Carlos B. Westphall
Yong Zhang
Xiaoliang Zhao
Wolfgang Zimmer
Simon Znaty
Douglas Zuckerman

Table of Contents

Architectures for Internet Management

Fault Management of Networks and Services

Management of Internet Services

Inter-domain Management

Event Handling for Management Services

QoS Management

Topics in Management Architectures

A Scalable Architecture for Monitoring and Visualizing Multicast Statistics

Prashant Rajvaidya[1], Kevin C. Almeroth[1], and Kim Claffy[2]

[1] Department of Computer Science
University of California–Santa Barbara
{prash,almeroth}@cs.ucsb.edu
[2] Cooperative Association for Internet Data Analysis
University of California–San Diego
kc@caida.org

Abstract. An understanding of certain network functions is critical for successful network management. Managers must have insight into network topology, protocol performance and fault detection/isolation. The ability to obtain such insight is even more critical when trying to support evolving technologies. Multicast is one example of a new network layer technology and is the focus of this paper. For multicast, the pace of change is rapid, modifications to routing mechanisms are frequent, and faults are common. In this paper we describe a tool, called Mantra, we have developed to monitor multicast. Mantra collects, analyzes, and visualizes network-layer (routing and topology) data about the global multicast infrastructure. The two most important functions of Mantra are: (1) monitoring multicast networks on a global scale; and (2) presenting results in the form of intuitive visualizations.

1 Introduction

Several useful network monitoring mechanisms have evolved over the years to support operational debugging and troubleshooting. The Internet Control Message Protocol (ICMP) and the Simple Network Management Protocol (SNMP)[1] are the original control and management protocols of the TCP/IP protocol suite. They form the basis for many monitoring tools. Despite these developments, monitoring the global Internet is still a formidable task. Its ever-increasing size and heterogeneity show the scalability weaknesses of existing management solutions. Generically, we believe that the basic challenges in global monitoring include: collection of data from a variety of networks; aggregation of these heterogeneous data sets; useful data mining of these data sets; and presentation of results. These challenges are applicable to almost any global monitoring system, and are applicable for almost any kind of data collected.

Recent network technologies such as multicast and Quality-of-Service (QoS) impose new requirements for network monitoring. Current deployment of these technologies in the infrastructure is far less than that of traditional unicast delivery. Because of the rapid pace of development, the lack of standards, and the lack of widespread understanding of these new technologies, the challenges for developing systems to monitor next-generation networks is a very difficult problem.

A. Ambler, S.B. Calo, and G. Kar (Eds.): DSOM 2000, LNCS 1960, pp. 1–12, 2000.

In this paper, we focus on multicast monitoring. Multicast provides a scalable and bandwidth-conserving solution for one-to-many and many-to-many delivery of packets in the Internet. Delivery of high-bandwidth streaming media via multicast not only improves the scalability of the streaming server (i.e., allows it to serve more clients) but also reduces the number of redundant data streams. Multicast was first widely deployed in 1992. Since then, the multicast infrastructure has transitioned from an experimental tunnel-based (virtual overlay) architecture based on the Distance Vector Multicast Routing Protocol (DVMRP)[2] to pervasive deployment of native multicast. In the current infrastructure, stable Internet multicast relies on a complex system of protocols operating in harmony: legacy DVMRP, Protocol Independent Multicast[3], the Multicast Border Gateway Protocol (MBGP)[4] for policy-based route exchange, and the Multicast Source Discovery Protocol (MSDP)[5] for exchanging information about active sources. Increased commercial interest and associated growth in multicast deployment makes monitoring both more important and more difficult. Systems that can gauge the performance of various multicast protocols, delineate various aspects of current multicast infrastructure, and predict future trends in workload are of tremendous value.

Our goal is to design and develop a system to monitor multicast on a global scale by collecting data at the network layer. We aim to use monitoring results to provide intuitive views of the multicast infrastructure. In this paper, we present Mantra, a tool that we have developed for this purpose. Mantra collects network-layer data by capturing internal tables from several multicast routers. Data is processed to depict global and localized views of the multicast infrastructure. Presentation mechanisms include topological and geographic network visualizations and interactive graphs of various statistics. Results from Mantra are useful for several purposes including assessing the amount of network activity, evaluating routing stability, and detecting and diagnosing problems. Another important feature of Mantra is its scalable and flexible architecture. Mantra provides mechanisms to easily support growth in the network as well as support for new data collection activities.

The rest of this paper is organized as follows. We review related work in Section 2. In Section 3, we describe goals challenges. Section 4 describes the design and architecture of Mantra. Section 5 provides an example of Mantra being used to identify network problems. The paper is concluded in Section 6.

2   Related Work

Monitoring the current Internet infrastructure on a global scale is challenging because it consists of a complex topology of numerous heterogeneous networks. Moreover, there is little interest for commercial Internet Service Providers (ISPs) to provide monitoring data to external organizations. Nevertheless, there is an array of useful work for monitoring the Internet beyond a single administrative domain. The earliest such tools include *traceroute* and *ping*. There are also several ongoing efforts in the field of end-to-end Internet monitoring, most involving active probe traffic sent from a source to one or several hosts and subsequent evaluation of response time, throughput, or path changes. However, most end-to-end monitoring tools and related analysis efforts lack intuitive visualization of results. As a consequence, proper interpretation requires an in-depth knowledge of the infrastructure and protocol operation. Tools also tend to be less than sufficient for detailed problem identification, isolation, and resolution.

A second challenge of monitoring the Internet beyond the complexities of the topology is the difficulty of monitoring multicast traffic. The difficulty arises primarily because of the differences between the unicast and multicast service models. In unicast networks, data transfer is between only two hosts. In contrast, in multicast networks, data is delivered to logical groups of hosts and data transfer takes place via a dynamic distribution tree. Consequently, monitoring multicast usually involves monitoring either the whole or a part of such distribution trees. In addition, a multicast sender does not typically know about all of a group's receivers. Therefore, even monitoring at the source is not straightforward.

The differences between unicast and multicast also reduce the effectiveness of using existing unicast monitoring mechanisms for multicast. In general, unicast tools provide only limited functionality and do not perform well for multicast-related network management tasks like data collection, data processing, presentation of results and provision for analysis. The solution has been to ignore existing unicast tools and develop new tools specifically for multicast.

There are a number of monitoring tools that have been developed specifically for multicast. One of the most widely used examples is *mtrace*[6]. It is an end-to-end tool that characterizes multicast paths between hosts. *MHealth*[7] provides a useful visualization front-end for *mtrace*, and MantaRay[8] attempted to do the same for tunnel information. However, both *mtrace*, and necessarily *MHealth*, suffer from scalability problems. The primary problem is that *mtrace* provides only a source-to-receiver trace and must be repeated for each group member. Large groups require large numbers of traces. Other tools, such as *mstat*, *mrtree*, and *mview*[9], collect data directly from routers via SNMP[1]. The limitation with SNMP-based tools is that they are typically only useful for intra-domain monitoring. Still another class of monitoring tools, including *mlisten*[10], *rtpmon*[11] and *sdr-monitor*[12], collect data at the application layer. While these tools provide important results, they provide little information about the network, router state, and network protocol operation.

3   Goals and Challenges

Monitoring multicast networks on a global scale requires mechanisms for collecting, analyzing, and presenting results. In this section we describe both our goals and the challenges of meeting these goals. We specifically frame this discussion in the context of Mantra but believe that our experiences are applicable to the construction of other, similar tools.

3.1   Goals

Design goals pertain to Mantra's architecture for data collection and analysis; presentation goals reflect the need to provide intuitive and useful visualization.

Design Goals. We have attempted to develop an appropriate generic architecture for collecting and processing data from multiple networks. Figure 1 depicts a simple model and the necessary stages. We need a flexible and scalable architecture for performing a wide range of monitoring tasks. As shown in the model, monitoring involves both data collection and data processing. Mantra's data collection occurs at the network layer, acquiring memory tables from multicast routers that are geographically and topologically dispersed throughout the world. Data processing requirements include: removing noise from raw data; converting raw data to Mantra's local data format; aggregating data collected from different networks; and analyzing these data sets to generate useful results. We elaborate on these tasks in later sections.

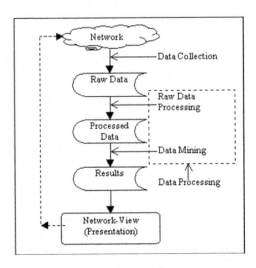

Fig. 1. Network Monitoring Model.

We also need Mantra's architecture to be flexible enough to accommodate the rapidly evolving multicast infrastructure. Frequent changes are common and may require modifications to the monitoring process. In addition, Mantra needs to be able to adapt to potential variations in the monitored environment, e.g., inconsistent raw data formats, unreliable data sources, and an unstable topology. Finally, Mantra needs to be able to handle a large, and increasing, volume of data; an inevitable consequence of the growing number of networks and protocols, as well as increased use in currently monitored networks.

Presentation Goals. We need to use collected and processed data to generate useful views of various aspects of multicast. We visualize results using several tools: Otter[13], for interactive topology visualizations; GeoPlot[14], for visualization of the geographic placement of various multicast entities; and MultiChart, a tool we have developed for interactive graphing. These mechanisms add to Mantra's usefulness for tasks such as: measuring performance of networks; estimating the extent of multicast deployment; debugging protocol implementations; detecting faults; identifying problems spots; and planning growth in the multicast infrastructure.

3.2   Challenges

As multicast has grown, so have the challenges associated with each step of the monitoring process. Some of the specific challenges include:

Challenges in Data Collection. Data collection from multiple sites poses a number of problems. The two most important are temporal variations in the allowed frequency of data collection and data format incompatibility. First, data collection is an invasive activity and will always add overhead to the router being polled. In the worst case, this additional overhead might contribute to overload, causing congestion and possibly the failure to handle the current traffic load. Second, different routers may be from different vendors and even routers from the same vendor will likely be running different versions of routing code. Each difference will likely affect the format of the data. Although protocols like SNMP exist to standardize the process of data collection as well as the format of collected data, there is a lack of SNMP support for multicast. Management Information Bases (MIBs) for the newer multicast protocols either do not exist or are not up to date. Consequently, SNMP is not suitable for monitoring newer multicast routing protocols.

Challenges in Data Processing. Data processing involves parsing raw data into well-structured tables and removing various types of errors from these tables. The first task requires keeping the parsing modules current with changes in raw data formats. The second task, error reduction/elimination, is extremely difficult to automate. Data can be noisy and unrepresentative of the true picture for several reasons, including: effect of test users joining and leaving sessions very quickly; incorrect data due to bugs in protocol implementations; and corrupt data because of problems during collection. Mechanisms to mitigate the effects of errors vary with the cause of the problem. While removing noise due to experimental user behavior involves developing heuristics to identify anomalies in data sets, managing data corruption might involve ignoring the entire data set.

Challenges in Data Mining. Challenges in data mining involve keeping our analysis techniques current with the rapid pace of multicast technology developments, as well as generating a representative global view of the multicast infrastructure. Problems with generating a global view are two-fold: (1) protocols such as PIM-SM and MBGP do not keep detailed global information, instead, they keep hierarchical information, i.e. they only keep information about reaching a domain and not how to reach hosts within the domain; (2) the lack of sufficient world-wide monitoring locations, data format compatibility, and temporal congruity makes it difficult to develop a consistent global view.

4   Design of Mantra

Mantra's architecture follows the basic model introduced in Section 3. Figure 2 depicts the information flow at different stages–from data collection to data processing, analysis and storage of results. We classify different entities that constitute this model into two broad categories: information (data) formats and module groups. In this section we describe these two categories in further detail.

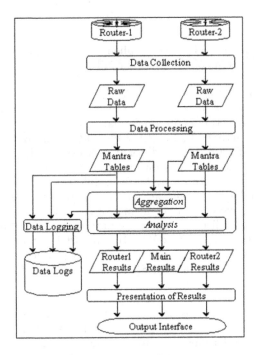

Fig. 2. Architecture of Mantra.

4.1   Information Formats

At any stage of a monitoring cycle, data can belong to one of the following three classes: intermediate results, data logs, or monitoring results. Intermediate results refer to the transient information passed on from one module-group to another during different stages of processing. Data logs refer to the final form of the data. These data sets are archived and used for future analysis. Monitoring results refer to the data that has been prepared for use as input for the visualization tools.

We have designed a set of tables, referred to as *mantra-tables*, which provide a standard framework for formatting different types of monitoring information collected from various sources. The two main benefits of such a framework:

- Analysis and aggregation modules remain transparent to the different raw data formats. We can make Mantra adopt to such changes simply by modifying the existing data processing modules or creating new ones. This process is further explained later in the section.
- Efficient data aggregation provides scalability by reducing processing requirements. It also facilitates a more accurate global view of various aspects of multicast by having a more consistent data set.

Based on their key data field(s), mantra-tables can be classified into two types: base tables and composite tables. Base tables hold information about the characteristics of basic multicast entities: groups, hosts, networks and Autonomous Systems (ASes). Composite tables hold data from multiple base tables, related to either the state of different protocols, or multicast routes. While

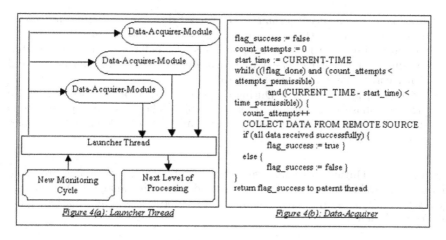

Figure 4(a): Launcher Thread

Figure 4(b): Data-Acquirer

Fig. 3. Data Collection.

Mantra uses the original route tables for archival purposes, it uses aggregated route tables for analyzing the routing data.

4.2   Module-Groups: Mantra Tasks

We divide Mantra functionality into four phases, each with a module group that performs the corresponding task. These four phases are represented in Figure 2 and each is discussed below.

Data Collection. Data collection involves capturing state tables from multicast routers. As mentioned above due to lack of updated standards, SNMP data can not yet be used for monitoring newer multicast protocols. Consequently, Mantra obtains router state by logging into the routers and capturing router's memory tables directly. The module group for data collection in Mantra constitutes of two modules: the launcher-thread and the data-acquirer. The launcher-thread initiates data collection from routers and passes the data to the next phase of operations; the data-acquirer module is responsible for the actual data capture. At the start of each monitoring cycle, the launcher-thread starts multiple instances of the data-acquirer module and then waits for all of them to finish before passing the data to the next module group. Collection from multiple routers thus occurs in parallel. This not only reduces the overall time required for collection but also increases the temporal vicinity of data from different sources. Figure 3(a) illustrates the launcher thread.

Raw Data Processing. Data processing consists of converting raw data captured from external sources to mantra-tables. We have developed a conversion module for each type of data set collected. These modules act as plug-in parsers for converting associated data types to appropriate mantra-table(s). Using separate modules for different data types makes Mantra easily adaptable to changes in formats. New parsers can quickly and easily be substituted for existing ones. The level of processing in these modules varies. Two important tasks that these modules perform are:

- *Rectifying Erroneous Information*: Collected data can be erroneous and/or unrepresentative of the true picture for several reasons, including: implementation bugs in the routers; anomalous user behavior; and incompatibility among adjacent routers. We need to detect inaccurate information and either correct or remove erroneous values.
- *Generating Mantra Tables*: During this stage, raw data modified during the previous phase is converted to mantra-tables. The conversion procedure is straightforward, and is typically a simple mapping of fields from raw tables to mantra-tables.

Data Logging. During this phase we archive mantra-tables containing processed data. These archives can later be used for in-depth offline analysis. Our primary goal is to minimize storage space requirements without loss of information. Techniques used include:

- *Storing Only the Deltas* : Mantra stores only the entries that have been either withdrawn or added since the last monitoring cycle. This technique is very useful for storing MBGP or DVMRP tables; tables that do not change often.
- *Utilizing the Relational Nature* : Many mantra tables can be grouped into sets such that combining tables yields data on some important entity. In some cases, such as when the primary key constitutes most of the information in the table, we merge tables into a single table and store only that table.
- *Splitting the Tables* : The opposite of joining tables is also a useful technique. Mantra may split a composite table into constituent base tables for archival. For example, we may split an mroute table into two tables: the sources table and the groups table. The advantage of table-splitting is increased ability to store deltas, since the possibility of temporal consistency between base tables is higher.

Data Analysis and Aggregation. During this phase, Mantra further processes data for analysis. Some aspects of multicast that Mantra analyzes include membership patterns, usage of multicast address space, MSDP performance, routing stability, host characteristics, and network characteristics. The format of these results is optimized for use with different output interfaces. For example, Mantra stores results from group size analysis in simple tabular format-primarily useful for graphing. Other results represent topology trees and are stored for use in topology visualizations.

Mantra also performs two types of data aggregation during this phase: (1) aggregation of various types of data sets; and (2) aggregation of similar data sets from different sources. The first type of aggregation allows us to broaden the scope of monitoring beyond the analysis of individual protocols. For example, consider the case of MBGP and MSDP. Both tables are monitored individually by Mantra, but which are often needed together, e.g., to assess propagation of Source Active (SA) messages or density of MSDP sources in MBGP domains. The second type of aggregation is critical to obtaining a global picture of the infrastructure and relating various types of data.

5   Presentation of Mantra Results

We use a set of static as well as interactive visualization mechanisms for presenting results. The types of results Mantra can produce support both a cursory examination of multicast statistics as well as detailed analysis of routing problems. In general, they allow study of multicast deployment, traffic load, protocol performance, and fault detection/isolation. In this section we describe these visualization mechanisms and demonstrate their utility with a case study of Mantra's use in detecting and isolating a routing problem.

5.1   Visualization Mechanisms

Mantra uses five output interfaces for presentation of results: (1) tables, (2) static graphs, (3) interactive graphs, (4) interactive topology maps and (5) interactive geographical representations. Of these, the interactive presentations offer important functionality and flexibility. We describe these interactive interfaces below and then present a case study using these interfaces in the next section.

Topology Maps. These provide graphical illustrations of different MBGP topology views. Mantra uses a Java-based, interactive topology visualization tool, Otter, for this purpose. Two types of views are: local views–the MBGP topology as seen from an individual router, and a global view–the MBGP topology obtained by aggregating data from different routers. Otter provides functionality through which user can interactively customize the colors of links and nodes based on values associated with them. Mantra can display statistics about various characteristics, including: node degree, link traffic, MSDP statistics, and distribution of participant hosts across administrative systems (ASes).

Geographic Placements. Placement provides a mapping of various components of the multicast infrastructure according to geographic location. Mantra uses the interactive Java-based tool, GeoPlot, to provide geographic placement of MBGP networks, DVMRP networks, participant hosts and RPs on a world map.

Interactive Graphs. Statistics are presented in the form of customizable graphs, using the MultiChart tool that we developed for Mantra. MultiChart provides a user-friendly interface for controlling different visualization aspects of the graphs, e.g., overlaying different graphs on the same display, choosing temporal range of data, and scaling graphs.

5.2   Isolating an Outage: A Case Study

In this section we present a case study of the use of Mantra to detect a routing problem, discover its cause, and evaluate its effects. The case we present pertains to a MBGP routing problem that we noticed on August 21, 1999 at ORIX, one of the routers that we collect data from. Below we present a step-by-step analysis.

Observation–The Unusual Results. Figure 4 (left graph) shows the number of session participants graphed over time. The point of this plot is the unusual drop in the number of sources at 1:56 am on August 21, 1999–the number of sources dropped by 23%. Such a severe and sudden drop is unlikely to be normal user behavior. It is likely the result of a routing problem.

 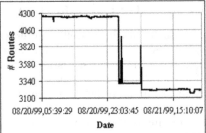

Fig. 4. Number of Participants (Left Graph) and MBGP Routes (Right Graph).

Problem Solving. MBGP routing statistics derived from the data collected in the same time frame confirm that a routing problem occurred. Figure 4 (right graph) shows the distribution of the number of MBGP routes as seen from ORIX. Here we noticed a sharp drop, about 22.2%, in the number of MBGP routes in the snapshot taken at 1:56am on August 21, 1999. This drop correlates with the number of participants (shown in the left graph of Figure 4).

The number of routes in a router's MBGP table should typically remain relatively constant, so a large change is a strong indication of a potential routing problem. However, it is difficult to derive an exact correlation between the loss of MBGP routes and a decrease in the number of participants. Other factors may conspire to make drops caused by a single event look less synchronized. For example, a large number of joins in another part of the topology may minimize the perceived impact. Our efforts to visualize MBGP topology help to provide additional data for verifying outages. Figure 5 (left graph) shows a screen shot of two consecutive snapshots of the MBGP topology overlaid on the same display. Links common to both topology snapshots are in light gray; those seen only in the second snapshot are black. The figure shows that an entire portion of the multicast infrastructure reachable via AS-704 is absent from the second snapshot.

Analysis of the Effects of the Problem. A detailed offline analysis showed that AS-704 provides links to several networks in Europe. Consequentially, loss in connectivity for AS-704 resulted in lost connectivity to most European networks. This confirms the loss in participant-hosts shown in Figure 5 (left graph). Our efforts to place participants on a geographical map offers another useful result. Figure 5 (right graph) shows geographic placement of participant hosts on a world map for both before and after. Figure 5 (top right graph) displays the hosts present before the drop, Figure 5 (bottom right graph) depicts the scenario after the drop. The difference in the density of the hosts in Europe between the two figures confirms the loss of connectivity to the countries Germany (.de), Czech Republic (.cz), and Greece (.gr).

6   Conclusions

Mechanisms for monitoring the Internet infrastructure on a global scale hold great value. However, developing such mechanisms is challenging due to the relentless growth in deployment, heterogeneity among networks, fast pace of developments, and lack of support for inter-domain monitoring. Current monitoring

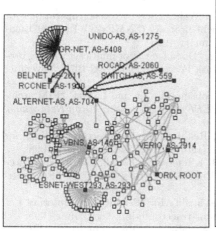

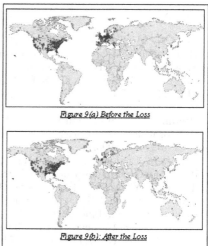

Figure 9 (a) Before the Loss

Figure 9 (b): After the Loss

Fig. 5. Loss in MBGP Connectivity (Left Graph) and Affect on Hosts (Right Graph).

systems provide only limited functionality, and are only marginally successful at intuitive visualization of results. With the emergence of the next generation of networking technologies, the need for new types of monitoring mechanisms has become urgent. Multicast is one such rapidly growing networking technology that requires effective monitoring to promote deployment and stable evolution. However, progress in multicast monitoring is hindered by several factors, including rapid changes in the field, incompatible standards, routing instability, and bugs in protocol implementations.

We have introduced Mantra, a tool developed for monitoring multicast on a global scale. Mantra collects network-layer data by capturing internal memory tables from routers across topologically and geographically diverse networks. Through Mantra we have developed a useful system for analyzing multicast behavior, including session characteristics, membership patterns, routing stability and MSDP performance. We have designed Mantra to be flexible; by keeping different modules independent of each other and by defining a standard data format for information flow amongst them we have created a model that can sustain intensive processing even as the number of networks and volume of monitored data grows. This model also enhances the scalability of Mantra as the processing of data sets can be easily distributed amongst different hosts or can be done at the source itself. Processed data can then be aggregated hierarchically and results can be generated based on a global snapshot.

We have described the visualization of monitoring results from Mantra with tools for interactive graphing of various statistics, topology visualizations, and geographic placement of different multicast subnets. We have also described how realtime results from Mantra can be used for gauging the current state of multicast and longer term results can be used for detecting faults, and discovering the cause of these faults. Finally, we have provided a case study to illustrate the utility of Mantra in troubleshooting a routing problem.

References

1. J. Case, K. McCloghrie, M. Rose, and S. Waldbusser, "Protocol operations for version 2 of the simple network management protocol (SNMPv2)." Internet Engineering Task Force (IETF), RFC 1905, January 1996.
2. D. Waitzman, C. Partridge, and S. Deering, "Distance vector multicast routing protocol (DVMRP)." Internet Engineering Task Force (IETF), RFC 1075, November 1988.
3. S. Deering, D. Estrin, D. Farinacci, V. Jacobson, G. Liu, and L. Wei, "PIM architecture for wide-area multicast routing," *IEEE/ACM Transactions on Networking*, pp. 153–162, Apr 1996.
4. T. Bates, R. Chandra, D. Katz, and Y. Rekhter, "Multiprotocol extensions for BGP-4." Internet Engineering Task Force (IETF), RFC 2283, February 1998.
5. D. Farinacci, Y. Rekhter, P. Lothberg, H. Kilmer, and J. Hall, "Multicast source discovery protocol (MSDP)." Internet Engineering Task Force (IETF), draft-mboned-msdp-*.txt, June 1998.
6. W. Fenner and S. Casner, "A 'traceroute' facility for IP multicast." Internet Engineering Task Force (IETF), draft-ietf-idmr-traceroute-ipm-*.txt, June 1999.
7. D. Makofske and K. Almeroth, "MHealth: A real-time graphical multicast monitoring tool for the MBone," in *Workshop on Network and Operating System Support for Digital Audio and Video (NOSSDAV)*, (Basking Ridge, New Jersey, USA), June 1999.
8. B. Huffaker, K. Claffy, and E. Nemeth, "Tools to visualize the internet multicast backbone," in *Proceedings of INET '99*, (San Jose, California, USA), June 1999.
9. *Merit SNMP-Based MBone Management Project.* http://www.merit.edu/net-research/mbone/.index.html.
10. K. Almeroth, *Multicast Group Membership Collection Tool (mlisten).* Georgia Institute of Technology, September 1996. Available from http://www.cc.gatech.edu/computing/Telecomm/mbone/.
11. A. Swan and D. Bacher, *rtpmon 1.0a7.* University of California at Berkeley, January 1997. Available from ftp://mm-ftp.cs.berkeley.edu/pub/rtpmon/.
12. K. Sarac and K. Almeroth, "Monitoring reachability in the global multicast infrastructure," in *International Conference on Network Protocols (ICNP)*, (Osaka, JAPAN), November 2000.
13. B. Huffaker, E. Nemeth, and K. Claffy, "Otter: A general-purpose network visualization tool.," in *INET*, (San Jose, California, USA), June 1999.
14. R. Periakaruppan, *GeoPlot - A general purpose geographical visualization tool.* Available from http://www.caida.org/Tools/GeoPlot/.

A Control Architecture for Lightweight Virtual Networks

Sean Rooney

IBM Zurich Research Laboratory
8803 Rüschlikon, Säumerstrasse 4, Switzerland
sro@zurich.ibm.com

Abstract. As an IP network is supportable over multiple different layer 2 networks, so a virtual network is supportable over multiple different resource allocation mechanisms. This fact is obscured in the literature as the goal — the privileging of certain user traffic by segregating it from other traffic — is tightly bound with the means of achieving it — e.g., the differentiation of traffic at routers based on the bits in the type of service field of the IP packet. We propose an IP control plane capable of supporting the dynamic creation of lightweight virtual networks and requiring minimal support from network devices. The latter is achieved by minimizing what is required from network elements, while taking advantage of what is available.

1 Introduction

At a fundamental level network policy is an emergent property of differentiating between traffic. Traffic differentiation can be distinguished based on the following three principles:

- the reason for doing it, e.g. QoS, security;
- the granularity at which it is performed; e.g. flow, user, service, site;
- the scope, e.g. between hosts, between edge routers.

Existing solutions address different parts of the problem space defined by this tuple. For example what is commonly called in the literature a Virtual Private Network (VPN), such as the commercial offer proposed in [1], secures traffic in transit between trusted locations; guaranteeing resources between hosts is addressed by the IETF integrated service model (int-serv) [2]; while the IETF differentiated service model (diff-serv) [3] allows a guarantee to be given about the forwarding of the aggregate of traffic across a router for a given service.

The diversity of ways in which policy is defined for, distributed to and applied on different network devices makes the configuration of the network complex and reasoning about its overall behavior extremely difficult; this is likely to inhibit the introduction of new desirable policies. We propose a single abstraction within which different solutions at different protocol layers can be coordinated. This abstraction interconnects applications resident on various edge elements such as

A. Ambler, S.B. Calo, and G. Kar (Eds.): DSOM 2000, LNCS 1960, pp. 13–24, 2000.

servers, edge-routers and hosts, in such a way that their communication can be distinguished from other traffic carried over the same physical infrastructure. As the abstraction isolates traffic from one user group from others it effectively virtualizes the physical network and is properly termed a virtual network. The qualifier *lightweight* is used to denote that these virtual networks are created quickly and make minimal requirements from the physical network over which they are overlayed.

Lightweight Virtual Networks (LVN) are currently being used for supporting network policy within a framework for the dynamic addition of new content providers to an Application Service Provider's (ASP) infrastructure [4]. The LVN is the means by which the ASP's distribution network is partitioned between the diverse clients and the location at which QoS guarantees are enforced.

We motivate the feasibility of the approach by describing a working implementation of a control architecture for creating LVNs.

2   LVN Control Plane Requirements

A Lightweight Virtual Network (LVN) is an overlay network allowing a community of information producers and consumers to exchange information such that at the network nodes their traffic can be distinguished from other traffic and can be treated according to some community specific policy. The constraints on the LVN control plane are such that it must:

- *potentially be end-to-end*: often policy, for example security, needs to be enforced everywhere or not at all.
- *involve both layer 2 and 3 devices*: the distinction between routers and switches is becoming blurred. Routers are forwarding at or near switching speeds, while Ethernet switches, for example, are becoming more sophisticated in regard to the policy they can support.
- *make use of existing technology and run in a heterogenous environment*: a successful solution must build on the large range of existing technologies for supporting diverse types of policy.
- *make minimal requirements of the network nodes*: while use will be made of any technology for supporting policy resident at a node, the solution must make as few demands as possible on the network nodes. Equipment as diverse as a simple Ethernet hub and an MPLS Router should fit into the same framework.
- *be dynamic and self sustaining*: LVNs are created dynamically without the intervention of a human operator.

The policy to be applied to traffic on a given LVN must be distributed to all appropriate nodes along with a label. Each data unit belonging to the traffic of that LVN carries the label enabling the node to identify the appropriate policy to use when forwarding the unit.

Network policy is the coordination of the diverse forwarding functions of a set of network devices to attain some overall objective, for example security through packet filtering, guaranteed service quality through priority based forwarding. No attempt is made to define policy precisely as this will necessarily evolve as network elements obtain new capabilities.

A LVN label can be something implicit, e.g. an IP source address, or explicit, e.g. an MPLS label [5]. The label may be present only at the IP layer or also be usable by layer 2 devices such as Ethernet and ATM. It may have significance for the network as whole, e.g. a VLAN [6] identifier, or have only local significance, e.g. an ATM VCI. Using implicit labels for identifying policy does not require modifying the format of the data units. They can therefore be carried by nodes that do not support the LVN infrastructure. However, they are more restrictive as they simply use parts of the format of the data unit for a purpose for which they were not intended. Explicit labels can be used exclusively for the purpose of identifying policy, but require changing formats of well established data units (or creating new encapsulation for them) with all the inherent problems. Moreover, there is no general agreement about what such a label should be like, e.g. MPLS, IPv6.

2.1   Design Choice

The chosen solution for the infrastructure is to use implicit labeling; no changes are required in the format of data units and no modification in the forwarding function of the nodes are required.

As the LVNs are supported both end-to-end and across multiple network layers, the implicit label chosen has to be meaningful to a large range of different network devices. The TCP/IP protocol suite is sometimes described in terms of an hour glass shape: many different protocols exist above and below the IP layer, but are unified at the IP layer. Although, in theory, layer 2 protocols should be oblivious to IP addressing, in practice they are often IP aware. For example an Ethernet switch may limit a broadcast to a VLAN containing only hosts within a given IP subnet. It achieves this by maintaining a MAC address/IP address mapping; this mapping is obtained by sniffing ARP requests and replies. If this type of sniffing is not supported then there is still the potential to use policy to enhance the IP address/Physical Address resolution, for example, a given set of IP addresses are resolved to both an Ethernet MAC address *and* a specific VLAN priority tag.

IP's ubiquity makes the IP address a good candidate for the implicit label within the prototype implementation.

Multiple LVNs should be able to coexist on the same host and the presence of LVNs on a given host should not affect its normal operation. Therefore, the LVN is not associated with the IP address of the host itself, but rather with a dedicated *private* IP address; the host has one such private address per LVN. Including socket identifiers as well in the label allows the discrimination to be carried right to the application layer, e.g. two instances of the same application on the same host may communicate with a given server using different LVNs.

At the IP layer a LVN appears as a set of private IP subnets. These subnets are created dynamically across the physical network and the participating end-

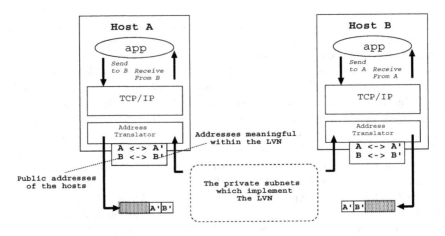

Fig. 1. Public/Private Mapping

points in the LVN are informed of the private-to-public mapping they should use in order that their traffic be carried across that LVN. The private IP address is the label which allows each node to determine the policy to use for forwarding that traffic. These private IP addresses are not exchanged by routers and therefore addresses conforming to RFC 1918 could be used, however as the LVN exists only over a well defined set of sites, any address space convenient for those sites is usable. Different administrative domains negotiate at LVN creation in order that the address space allocated to the LVN during its lifetime is used by it alone across all the involved domains. How this is achieved is explained in Section 3. The addresses used by a LVN are available for reuse after it is deleted from the network.

Figure 1 shows two hosts with addresses A and B participating in a LVN. First a private IP subnet (or set of subnets) is created across the nodes that interconnect the hosts. Each of the participating hosts receive mapping: A→A', B→B', such that A' and B' are meaningful addresses within the LVN's private subnets. The hosts update themselves such that if A wishes to communicate with B within the context of the LVN then A should send with source address A' and destination address B'. The network nodes identify the traffic within the LVN using the private address apply the LVN specific policy when forwarding it.

Within the dynamic ASP infrastructure described in [4] instead of mapping public-to-private addresses within the protocol stack, multiple virtual servers, each with their own private IP address are started on the ASP's physical servers on behalf of a content provider. Proxies with public DNS registered addresses running at the edge of the ASP's network are used by end-users to access the content; the proxies communicate with the appropriate virtual servers, and the LVN is used to distinguish the traffic of one content provider from another within the ASP's network.

3   LVN Admission Control

The creation of virtual overlays requires coordination of the participating entities before creation can occur, for example keys must be exchanged if encryption is required, pricing negotiated agreed upon, etc. This is especially true when multiple administrative domains are involved. In consequence, the authorization function is split into two: first, the application communicates to an authorization entity its requirements for the virtual network and obtains from this entity the system wide information that participating nodes will need to know in order to create it; second, the application propagates this information to each of the network nodes which will then make a local decision to accept and forward or reject and rollback.

3.1   Design Choice

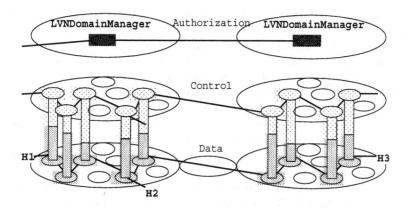

LVNDescription: [(H1,H2,H3) First Class]

LVNContract: [H1->H1', H2->H2', H3->H3', "First Policy"]

Fig. 2. Authorization

In the current implementation, the authorization entity is named the LVNDomainManager, the abstraction the application passes to this entity is the LVNDescription and the system wide state it returns the LVNContract. LVNDomainManagers are federated such that there is one per administrative domain and they intercommunicate in order to obtain the system wide information.

The LVNDescription is simply a set of IP addresses of the end-points of the intended LVN (in the case where the LVN is supported from application to application, socket identifiers are included as well) and a policy description.

The `LVNDomainManagers` collaborate together in order to resolve the `LVNDescription` into a `LVNContract` acceptable to all participants. In the current implementation the contract is a set of mapping of public addresses to private ones. More generally the contract would contain additional information such as encryption keys or different LVN label formats, e.g MPLS labels, VLAN identifiers, used to create the end-to-end LVN.

When a `LVNDomainManager` receives a `LVNDescription`, it determines which addresses are local to the domain and which foreign. For any foreign address the `LVNDomainManager` contacts (via a directory service) the appropriate `LVNDomainManager`. If no such entity exists then the LVN creation can still continue but the policy can only be applied within the local administrative domain. The `LVNDomainManager` must decide if this is appropriate. For example, in the current implementation the `LVNDomainManager` will continue the LVN creation but the public addresses for the foreign hosts will be mapped to themselves. Figure 2 shows the patterns of interactions between entities required for the creation of LVNs across multiple domains involving three hosts with addresses H1, H2 and H3.

4   LVN Adaptors

Both layer 2 and layer 3 devices can offer support in regard to supporting policy. For example an IP router might support diff-serv priority based queuing on bits in the IP header, an IEEE 802.1q [6] enabled Ethernet switch priority switching based on frame tags, and an ATM switch continuous bit rate cell forwarding based on the virtual circuit identifier. All might be required in order to give an end-to-end performance guarantee. However, the label, the attachment to the policy and the means of disseminating the association are all very different. In order to be as general as possible the infrastructure supporting the virtual networks should be able to handle multiple different technologies.

The principle adopted is that the underlying node should try to do as *much as it can* to support the virtual network, but its exact behavior is a function of the technology and the precise capabilities of the element. This is called the *principle of low expectations*. By adopting such a weak semantic a wide variety of technology can be coordinated within the same framework. The LVN control plane guarantees connectivity for the traffic of the LVN, but gives no precise guarantee as to how the associated policies are enforced. The behavior of the LVN control plane may be thought of as a type of 'best-effort policy support', it only guarantees not to make the situation worse for supporting the policy on its associated network elements, it does not guarantee to make it any better.

4.1   Design Choice

In the prototype a `LVNController` entity is associated with each node within the physical network. The controllers support a common interface for LVN control independent of the nature of the underlying element. Within each controller is an adaptor that maps generic LVN operations onto a particular technology and then further onto a particular instance of that technology, i.e. the model of switch, router, etc. So for example, while a router and an Ethernet switch both support a `createLVN` operation taking a `LVNContract` as argument, their effect is very different: the router updates its routing tables to reach the newly created private IP subnets with an appropriate forwarding class, while the Ethernet switch might support the LVN through the creation of a IP rule based VLAN with associated priority.

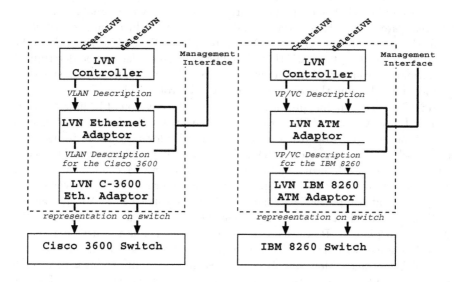

Fig. 3. Example of Adaptors

Figure 3 shows an example of two adaptors for two different technologies and pieces of equipment.

The `LVNController` may be thought of as a convenient abstraction in which to collect together a variety of architectural entities currently supported by network elements. For example, RSVP [7], COPS [8] the MPLS Label Switching Router [5] etc., would all be constitutions of the controller. The controller supplies an interface for LVN control general enough and with weak enough associated semantics such that it is supportable over a range of technologies, but the actual implementation of the operations is performed using existing technologies. The adapter is nothing other than a specification of how the mapping is performed between the `LVNController` interface and a network element with a given set of capabilities.

The exact nature of the mapping is set and adjusted by the network operator via management. For example, a new enterprise specific policy ('The-quarterly-report-broadcast-policy') can be added to an LDAP enabled directory along with the precise mappings to be used for a range of technologies. When the `LVNController` encounters an unknown policy in the `LVNContract` it communicates with the directory in order to obtain the appropriate mapping.

5   Propagating LVN Control Messages

To create an LVN, the entire set of specified edge elements and some set of network elements interconnecting them must be made aware of the policy and the associated label. The normal network is used to carry the signaling messages that instigate the creation of the virtual networks over which the data will flow. From the point of view of the LVN control plane, the public network is the signaling channel — the path that signaling messages take will be determined by normal IP routing. While it would be possible to define a new signaling protocol for achieving this task, reusing existing protocols is preferred.

RSVP [7] is a signaling protocol designed for the creation of IP flows. Adopting RSVP as the means of LVN creation has the advantage of using a well known and established protocol. RSVP soft state model allows the LVNs to have the desired self sustaining property, the use of IP routing to determine the topology of the virtual network means that support from the network elements is minimal and the fact that RSVP does not specify the precise form of the flow specification allows for arbitrary policy/label associations to be carried in RSVP messages.

However, RSVP as defined in [7], does not match exactly the needs of the LVN control plane as RSVP uses IP routing to determine the next hop of the control message, within the LVN control plane layer 2 devices must also receive the RSVP messages. Moreover, RSVP is designed for applications with a small number of senders and a large number of receivers, the advertisement of a flow (via a Path-Message) is performed by an information producer, and the creation of the flow initiated by an information receiver. In consequence RSVP is inherently asymmetric; the LVN described in this paper should satisfy a larger class of applications. For example, within a ASP traffic is carried both to and from the servers. The final problem is that the flows that RSVP creates are trees, with the sender as root and the receivers as leaves (there may be several senders in which case it is a forest). The LVN is (by definition) a graph.

5.1   Design Choice

In the prototype an enhanced RSVP daemon is run as part of the `LVNController`. In order to allow RSVP message to be propagated to both layer 2 network nodes and routers, the forwarding function of the daemon is a function of the nature of the device it controls. Routers use their routing tables in order to identify the next router to forward a message to in order to reach a given end-point. They then use physical topology information to determine which is the next layer 2 hop to take them to that router, the message is then forwarded to that layer 2 entity with the next router as subtarget.

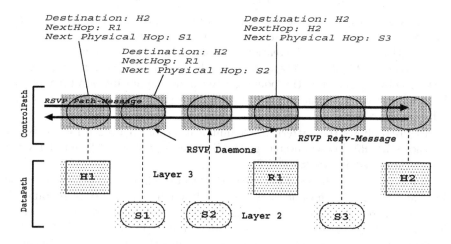

Fig. 4. LVN Creation Using RSVP Messages

The RSVP daemon of layer 2 entities only attempts to find the next hop to reach the subtarget rather than the end-point, they do this using physical topology information, so for example an Ethernet switch needs to know the next Ethernet switch to cross to get to an IP router. The view of the physical network that a LVNController is required to have is restricted to the boundaries of the set of layer 2 technologies to which it is attached. Layer 2 topology discovery is technology specific. For implementation purposes a simple Ethernet discovery system has been created. In this implementation each Ethernet LVNController registers its port/MAC association with a LVNDomainManager and the LVNDomainManager correlates the information to obtain the view of the entire Ethernet. This method could easily be replaced by appropriate layer 2 discovery mechanisms if available in the environment.

If a layer 2 network element is not discovered then it is transparent to the control path and no RSVP is forwarded to it. This does not mean that data for the LVN will not cross that network element, but only that no policy can be enforced on it. For example, if a host is connected to a router across an Ethernet hub and the hub does not support a LVNController then it is not discovered and the RSVP message from the host is forwarded directly to the router. This is simply another application of "the low expectations principle".

Among the set of end-points specified in the contract one is nominated as the initiator of the LVN creation. In the current implementation the element that requests the creation is considered the initiator, but another end-point might be chosen (perhaps based on its location in regard to all other end-points).

The initiating end-point starts by emitting RSVP Path-Messages, passing in the message the LVNContract. Each LVNController that receives a message, examines the contract to determine if it can support it and if so reserves the required resources. If it is the LVNController of a network node it then forwards the message to the next `LVNController` upstream, if on the other hand it is the `LVNController` of an end-point specified in the contract and it is willing to support the virtual network creation, it replies by sending reservation messages downstream. A `LVNController` receiving a first reservation message for a given virtual network activates the policy on the node, i.e. creates a VLAN on the Ethernet switch, a virtual circuit on an ATM switch and updates the routing table on the router.

Figure 4 shows the pattern of communication for creating a LVN between two hosts `H1`, `H2`, interconnected at the IP layer by a router `R1` and at layer 2 across three switches `S1`, `S2` and `S3`.

Implicitly a tree shaped LVN is created with the initiator as the root of the LVN and the others end-points as leaves. Although reachability is guaranteed, the route taken may be extremely inefficient, i.e. all communication between leaves must travel across their nearest common ancestor in the LVN tree. The `Path-Message` that an end-point receives contains the `LVNContract`, therefore the receiving end-point has access to the complete set of other end-points participating in the virtual network, allowing it to locally decide if there is a better route between itself and one or more of the other non-sender nodes rather than simply going through the sender. If it decides that a better route exists it starts sending out `Path-Messages` carrying the same contract as it received. The `LVNController` will not forward `LVNContracts` over interfaces where they are already supported; normal routing will ensure that the better route to the other hosts are included as part of the virtual network. In summary, a tree is created and then if better paths between the leaves exist, the leaves are joined in order to make a graph.

The LVN policy is soft state and is maintained by the periodic reception of RSVP messages carried in the public network. If the RSVP messages are carried along a different path due to routing changes, then the shape of the LVN is modified accordingly. This permits the LVNs to dynamically adapt to the current state of the network.

6   Related Work

Virtual Network Research: [9] introduced the notion of a *switchlet*, i.e. a switch partition that allows multiple control architectures to be supported over the same physical switch, [10] describes the Tempest framework in which switchlets are used. Several other research group have made similar proposals [11,12]. That work is weak in its: applicability to technologies such as Ethernet, which have little support for traffic seperation; in the interoperation of multiple technologies. The work described here addresses these issues by using IP as the glue and applying a type of 'best-effort' partitioning mechanism. While the Tempest allows network operators to associate distinct control systems with their set of switchlets, the LVN is simply created using one of a set of predefined policies; i.e. it is a lowest common denominator. An LVN could be supported using switchlets and this would be the preferred adaptor if the environment supported it.

IP QoS: Diff-serv [3] allows IP routers to apply a certain QoS policy to an aggregate of traffic associated with a given service. Traffic at a router is distinguished using the Type Of Service (TOS) field in the IPv4 header. Diff-serv is normally only enabled at edge routers. The policy/label association is typically manually set by a network operator directly on the router or added to a directory and distributed to the routers via a directory access protocol. The work described in this paper differs from diff-serv in that it attempts to use more general policies than just those concerning QoS, associate them with other granularities than traffic aggregations, apply them potentially from application to application and do all this dynamically. Moreover, the work in this paper attempts to make *direct* use of the sophisticated support offered by layer 2 devices for supporting policy. However, the LVN control plane makes use of diff-serv if enabled at the routers within its domain.

MPLS Based VPNs: Much work has been done on using MPLS [5] to instrument VPNs. For example, [13] describes how MPLS tunnels can be used to support IP based VPN between MPLS enabled routers capable of supporting VPNs; it terms such routers VPN Border Routers (VBRs). The VBR implements a virtual router for each VPN they support which in its turn services a separate forwarding table. The VBR forwards traffic from an enterprise across appropriate MPLS tunnels to a remote VBR at the appropriate site. VBRs are configured such that they have an interface address in the enterprises address range. VBRs for the same enterprise discover each other and exchange routing information using normal routing protocols.

The architecture in [13] is complementary to the work described in this paper. Traffic from multiple LVNs could be aggregated over a single MPLS VPN, or a dedicated VPN could be assigned to a LVN. The LVN-RSVP messages would be carried transparently over the MPLS tunnels to the correct sites. To integrate the two approaches the VBR must be capable of supporting the dynamic addition of new private subnets to and from the enterprise.

7   Conclusion

The diverse support for policy across different network technologies and in different scopes makes implementing the policy end-to-end problematic. We have proposed a unifying abstraction — a lightweight virtual network — with which a set of different policies may be associated across a range of technologies. Such lightweight networks are created dynamically without the intervention of human operators and have minimal requirements for the policy support of the underlying physical network. The virtual network control plane takes advantages, if available, of any support, for example diff-serv, ATM P-NNI, offered by the underlying devices without requiring their presence. This paper has motivated the feasibility of the approach by describing an infrastructure which supports the creation and management of such lightweight virtual networks. Current work is using LVNs as part of an integrated approach in supporting resource guarantees within an Application Service Provider's distribution network.

References

1. Cisco, "Virtual Private Networks(VPN)," *Cisco System Inc. marketing information*, 1999.
2. R. Braden, D. Clark, and S. Shenker, " Integrated Services in the Internet Architecture: an Overview," *Internet RFC 1633*, June 1994.
3. D. Black, S. Blake, M. Carlson, E. Davies, Z. Wang, and W. Weiss, "An Architecture for Differentiated Service," *Internet RFC 2475*, May 1998.
4. S. Rooney, "The ICorpMaker, a Dynamic Infrastructure for ASPs," *Proceedings of IEEE Workshop on IP Operations & Management*, Sept 2000.
5. E. Rosen, A. Viswanathan, and R. Callon, "Multiprotocol Label Switching Architecture," *draft-ietf-mpls-arch-06.txt*, August 1999.
6. IEEE/ISO/IEC, "Virtual Bridged Local Area Networks," *ISO Publication*, July 1998. Draft Standard: IEEE Standard for Local and Metropolitian Area Networks, P802.1Q/D11.
7. L. Zhang, S. Deering, D. Estrin, S. Shenker, and D. Zappala, "RSVP: a new resource ReSerVation protocol," *IEEE Network*, vol. 7, pp. 8–18, September 1993.
8. J. Boyle, R. Cohen, D. Durham, S. Herzog, R. Rajan, and A. Sastry, "The COPS Common Open Policy Service Protocol," *Internet RFC 2748*, January 2000.
9. J. van der Merwe and I. Leslie, "Switchlets and Dynamic Virtual ATM Networks," in *Integrated Network Management V*, pp. 355–368, Chapman & Hall, May 1997.
10. S. Rooney, J. van der Merwe, S. Crosby, and I. Leslie, "The Tempest, a Framework for Safe, Resource Assured, Programmable Networks ," *IEEE Communications Magazine*, vol. 36, pp. 42–53, October 1998.
11. A. Campbell and al, "The Genesis Kernel: A Virtual networking operating system for spawning network architectures," *Openarch'99*, March 1999.
12. W. Ng, A. Jun, H. Chow, R. Boutaba, and A. Leon-Garcia, "MIBlets: A Pratical Approach to Virtual Network Management," in *Integrated Network Management VI*, IFIP & IEEE, Chapman & Hall, May 1999.
13. L. Casey, I. Cunningham, and R. Eros, "A Framework for IP Based Virtual Private Networks." IETF draft-casey-mpls-vp-00.txt, November 1998. Work in progress.

MRMA: A Multicast Resource Management Architecture for IP Platforms

Ryan Wu[1] and Irene Katzela[2]

[1] ECE Department, University of Toronto, Toronto, Canada
ryan.wu@utoronto.ca
[2] Lucent – Networkcare, Toronto, Canada
katzela@lucent.com

Abstract. Multicast with QoS guarantees is a bandwidth-efficient transmission scheme for the delivery of multi-user multimedia applications. Unlike most other related work, which focuses primarily on the algorithmic or protocol aspects of the multicast problem, our work focuses on the architecture and implementation aspects. In this paper, we present MRMA, a multicast resource management architecture designed to provision multicast with QoS guarantees over the IP platform. MRMA offers a flexible framework for integrating the necessary components for implementing the key multicast functions that are needed for creating, maintaining and terminating multicast sessions: QoS-sensitive route selection and tree construction, address allocation, and session advertisement. Taking into account network operations that are unique to multicast-with-QoS, MRMA implements a hierarchical architecture that features centralized monitoring of network state and semi-distributed per-session multicast management. The architecture also implements the idea of moving the heavy computational load of administering multicast operations outside the network and onto the hosts.

1   Introduction

The exponential growth of today's Internet is fueled in part by an emerging class of multi-user broadband applications such as video-on-demand, video conferencing, and file distribution. These applications are generally characterized by high transmission bandwidth, high data processing and stringent Quality-of-Service (QoS) requirements. Multicast with QoS guarantees, referred to in this paper simply as multicast-with-QoS or multicast, is an efficient transmission scheme for the delivery of multi-participant multimedia applications with respect to bandwidth resource usage. It emphasizes on minimizing the amount of traffic that is needed for distributing the data to its destinations.

Multicast has sparked deep interest among the research and industry community in the last ten years since Deering first described the standard multicast model for IP networks [1], and there are many research papers written on each of the key multicast topics/functions. We list some of them here. For multicast routing algorithms, there exist many proposed heuristics for the classic Steiner Tree and Constrained Steiner Tree problems [3]. For multicast protocols, there are two categories: QoS-oblivious –

A. Ambler, S.B. Calo, and G. Kar (Eds.): DSOM 2000, LNCS 1960, pp. 25 - 36, 2000.
© Springer-Verlag Berlin Heidelberg 2000

PIM, CBT, BGMP, DVRMP, MOSPF, and QoS-sensitive – YAM, QoSMIC[6]. For multicast address allocation, there is the suite of IETF protocols that support intra-domain and inter-domain address allocation: MASC, AAP, MADCAP, and MAAA. For session announcement, there is the *sdr* tool and SDP protocol. Lastly, for reliable transport service, there are many proposed multicast transport protocols such as SRM, RMTP-II, PGM, ARM. We would like to note that all the work listed above analyzes the multicast problem from either the algorithmic or protocol aspect, and generally focuses on a one or two topics. In particular, none of them approach the multicast problem as a whole and address multiple multicast topics together.

Our work takes an architectural approach to addressing the multicast problem as a whole. The goal of this paper is to propose an implementable solution that addresses all the key multicast functions. We introduce MRMA, a multicast resource management architecture that is customized for provisioning multicast-with-QoS service over the IP platform. MRMA is not simply a multicast protocol, but rather a management architecture that integrates all of the key multicast components into a single coherent system for managing the setup and configuration of multicast-with-QoS connections. The key multicast components are: i) a protocol for conducting QoS-sensitive route selection and tree construction, ii) an addressing scheme for managing the allocation of multicast addresses, iii) a session directory for announcing the presence of active multicast sessions to the network, iv) and a transport protocol for providing reliable delivery of multicast data.

The rest of this paper is organized as follows. Section 2 discusses some related work. Section 3 presents the MRMA architecture and discusses the key design issues. Section 4 suggests some future research directions for on the MRMA.

2   Related Work

In this section we present a brief description of MBone, a primitive best-effort multicast architecture. Note however, that it is only somewhat related to MRMA. We are currently not aware of any other proposed architecture that is exactly comparable to the MRMA, that is, a management architecture for provisioning multicast service with QoS guarantees.

Multicast Backbone (MBone). The MBone [2] is an experimental prototype for demonstrating the feasibility and cost-saving advantages (in terms of bandwidth) of the multicast scheme, and is expected to be eventually phased out. It is an over-lay network and utilizes the DVRMP and MOSPF protocols to handle the routing functions; the session directory tool, *sdr,* as the session announcement server; and manual address allocation. This is the only work that we are aware of that, similar to our work, strives to address all the key components necessary for enabling multicast over the Internet. However, unlike MRMA, it can only provision best-effort multicast service.

The fundamental problem is that the MBone model was constrained by the network infrastructure to essentially 'emulate' multicast service over a native unicast-only Internet. It focused on extending and adding onto the existing IP protocol suite, which is specialized for best-effort unicast connections. For instance, the DVRMP and MOSPF multicast routing protocols are in fact merely simple extensions of the unicast routing protocols RIP and OSPF respectively.   As such, DVRMP and MOSPF do not incorporate any of the QoS-sensitive multicast algorithms described in [3], and consequently the multicast trees they produce can only be expected to satisfy very simple tree constraints or optimizations.   Moreover, the MBone is also lacking a connection-admission-control mechanism to regulate access to the network, a resource reservation mechanism that acquires network resources on behalf of the multicast group and reclaims idle resources on behalf of the network, a session manager to track relevant group membership information, and an effective monitoring mechanism that can generate accurate, up-to-date topology information for routing and traffic control purposes – all of which are essential for providing multicast and QoS guarantees. The MRMA architecture, on the other hand, does not suffer the same drawbacks.

3   MRMA Overview

In this section we present the design for MRMA, a multicast management architecture customized for provisioning multicast-with-QoS over IP networks. The design takes into consideration the group-centric properties that are unique to multicast in order to maximize the scalability, flexibility and accuracy of the architecture.   Unfortunately, due to length restrictions on this paper we can only provide an overview of the architecture here.   More detailed specifications are documented in [5].   We begin the discussion by describing the operation context of MRMA: operating platform and provisioned services.   Next, we continue with a description of the architecture and discuss the key design decisions.   Finally, we show how the MRMA can be applied to provision an example multicast service, a layered video multicast application.

3.1   Context Model

As is the case with any network management architecture, a detailed implementation specification of the MRMA is highly dependent on its operational context.   In other words, the exact design for the architecture is governed by the context model which specifies the properties of the physical network and the services it supports. As such, in order to facilitate detailed specification of core MRMA features we narrow the scope of the current version of the management system.

We are designing the MRMA to operate over a 'next-generation' IP platform that contains primarily 'intelligent' routers with native support for both unicast and multicast routing, flow-recognition, and local control of CPU and link bandwidth allocation.   The routers are also programmable. Such routers eliminate the need to implement IP tunneling, and allows QoS-sensitive packet scheduling and dynamic control of network configuration.   We are also assuming each domain is small enough in terms of number of nodes and physical size such that it is feasible to implement some form of centralized monitoring mechanism without the associated congestion or bottleneck problem arising from too much signaling traffic converging at one point. MRMA supports one-to-many intra-domain multicast that adheres to the Express

Multicast Model [4], which makes fundamental changes to the traditional Deering IP multicast model [1] used in the MBone. Express is a single-source service model that allows only one source to multicast on each group. We would like to stress at this point that although our current MRMA version is customized for intra-domain operation, the general MRMA architecture can be extended to provision inter-domain operation. Suggestions for such extensions are presented in [5]. A detailed description of the MRMA context model is also presented in [5].

3.2   MRMA Architecture

From the functional perspective, the architecture can be viewed as comprising of various functional blocks that are 'contained' within a kernel, as illustrated in Fig. 1. We model each of the multicast functions route selection, address allocation, session advertisement and traffic and congestion control as Functional Modules (FMs). The FMs have well-defined input/output specifications and perform very specific functions. This is a feasible approach since each multicast function is unique, and so each FM can be modeled as independent blackboxes whose internal design and implementation do not have a major affect on the design of the rest of the management system as long as it produces the desired output given some input.

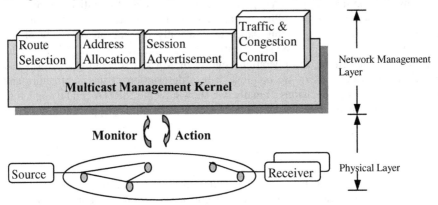

Fig. 1. Functional View of the MRMA Architecture

The Multicast Management Kernel (MMK) is the core component that integrates the various FMs into a coherent system. MMK coordinates the execution of high-level multicast operations such as creating a new session, adding a new member, etc. by invoking services from the FMs. In return, the kernel supports the FMs by performing housekeeping tasks such as gathering accurate network state information to be used as input for FM decision-making. MMK communicates with the FMs through well-defined interfaces.

The advantage of implementing each of the key multicast functions as modules is that it increases the adaptability of the MRMA architecture. It is designed such that the MMK will be able to swap-in and swap-out, with minimal customization, different solutions for each of the multicast functions. Some of the existing implementations and proposals for the key multicast functions listed in Section 1 can be modified to serve as FMs.

From an implementation perspective, the MMK and FM model discussed above is not implemented as a single software entity, but is rather implemented as a set of distributed, hierarchically-interacting management entities: Domain Manger (DM), Session Manager (SM), and Element Agent (EA). The collection of entities control network operation at three levels: there is one DM per domain that manages domain level operations; there is one SM per multicast session that manages session level operations, and there is one EA per network router that manages local node level operations. The entities share a hierarchical manager-agent relationship. One DM manages multiple SMs and EAs, and each SM in turn manages multiple EAs. The EAs manage the local node functions and also serve as agents to the DM and SMs. The function and design of the management entities reflect our desire to develop an architecture that effectively balances two generally conflicting management goals: maximize the distribution of the heavy processing load of managing each multicast session while ensuring the system is still capable of making accurate management decisions. These two goals are conflicting in that, generally, good load distribution is best achieved with a distributed management paradigm while accurate decision-making is better with a centralized management paradigm. The architecture for the DM and SM are illustrated in Fig. 2 and Fig. 3., and are described in detail in [5]. The architecture for the EA is not included in this paper, but it is also described in [5].

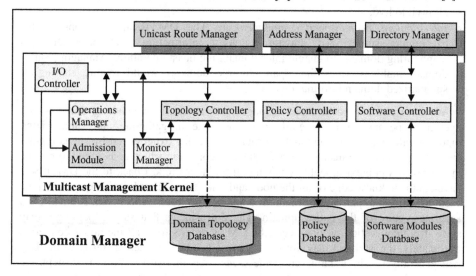

Fig. 2. Domain Manager Architecture

Domain Manager. The Domain Manager implements centralized management of domain-wide operations. The DM architecture is shown in Fig. 2. Some major functions of the DM are:

- *Topology tracking*. The primary DM function is to continuously monitor and maintain an accurate, up-to-date image of the domain state. The Monitor Manager uses a combination of explicit polling and 'trap-directed' polling [5] to collect node, link, traffic, and QoS metrics such as bandwidth usage and link congestion from the domain. The status information is then passed to the Topology Controller for storage in the Domain Topology database. The Monitor Manager ensures the collected information is accurate and up-to-date by taking steps to verify the integrity of received status messages. Damaged and out-of-sequence messages are filtered out. Topology image information is used as input for making route selection and admission control decisions.
- *Directory service*. Multicast session names and the corresponding session addresses, maximum source rate, and session duration are some of the properties tracked by the Directory Manager, a FM which serves a similar functionality as the *sdr* tool deployed on the MBone.
- *Address management*. The Address Manager is another FM which controls the allocation of a limited pool of multicast session addresses. It can implement a scheme based on the general Multicast Address Allocation Architecture (MAAA) .
- *Administer domain-wide policies*. DM acts as a policy server for controlling the distribution of policies and rules that govern domain-wide decisions. For instance, some domain-wide policies may limit the maximum aggregate bandwidth that a session is allowed to consume, or the maximum session duration that a session is allowed to live.
- *Inter-domain peer-to-peer communication*. The DM is also responsible for exchanging summarized domain topology information with peer DMs in neighboring domains to enable inter-domain multicast operation. Managing multi-domain multicast trees that span multiple domains would require sharing of summarized domain information.

We have chosen a centralized management paradigm for handling domain-wide operations because it is generally more accurate in gathering and summarizing topology information, as long as the propagation delay experienced by status reports from the agents to the manager is acceptable, a condition which is assumed to be true in the MRMA context model. By having all the nodes report back to the DM, it will have complete knowledge of all the node and link states. That knowledge can then be used to correlate partial information contained in each report and formulate an accurate picture of the entire domain. The downside is that such a paradigm may suffer from congestion on links near the DM as reports from all the nodes converge. Fortunately, the problem may not be a major since most signaling traffic to the DM will generally be short status report messages. An alternative option is to deploy a semi-distributed paradigm, whereby there are several DMs per domain, and each DM has complete knowledge of a portion of the domain. However, this method suffers from that fact that there is no single source from which other network entities can obtain a full, accurate, up-to-date image of the entire domain.

Since the DM represents a centralized point of convergence for a large volume of signaling traffic, its physical location in the network has an important impact on the performance of both the DM and the nearby network links. Therefore, the DM should reside in the 'center' of the network such that it maximizes the distribution of signaling traffic across the network and minimizes the maximum propagation delay experienced by status reports from network nodes [5].

Session Manager. To tailor the MRMA to handle requirements that are unique to multicast, we introduce the notion of a Session Manager. There is one SM dedicated for managing all operations relating to the setup and configuration of each session. SMs are dynamically spawned by the multicast sources to create and manage the multicast sessions. A SM resides on the same host computer as the source and shares the same IP address, but communicates on different ports. The SM architecture is shown in Fig. 3. The two major SM functions are:

- *Multicast tree construction.* Each SM manages the construction of a multicast tree that spans the source and all the group members. Tree construction involves initializing the session by registering with the DM, grafting new members onto the tree, pruning departing members and terminating the session. We will elaborate on these operations when we introduce our multicast protocol in Section 3.3. Multicast route selection and resource reservation are two important steps in tree construction.

   The SM implements a centralized route selection scheme. To add a new member, Rnew, to the group, the SM first invokes the Multicast Route Manager to attempt to compute a feasible route from the existing tree to Rnew. The Route Manager's multicast routing algorithm takes as input the current state of the entire domain, the current configuration of the tree, and the set of application or receiver-specific QoS metrics. The computed route, Path(Src, Rnew) must satisfy the QoS constraints defined by the application or Rnew, where Src is the multicast source. Note that a section of Path(Src, Rnew) coincides with the existing tree (*trunk section*) while the other section represents the new branch that needs to be grafted (*branch section*). QoSMIC also introduces the notion of a Manager Router for managing a centralized Multicast Tree Search Procedure [6].

   Other than selecting the path, SM is also responsible for reserving resources from all the on-path nodes along the Path(Src, Rnew). Along the *trunk section*, SM must ensure the on-path nodes reserve additional bandwidth, if required, to accommodate the QoS requirements of the new member. Similarly along the *branch section*, SM must ensure the nodes reserve sufficient bandwidth and add an entry for Rnew in their multicast routing tables. Once a feasible route is found and resource reservation negotiations are successfully completed, Rnew can begin receiving multicast traffic.

- *Multicast tree performance maintenance.* Similar to the DM, the SM also has a Monitor Manager that continuously monitors the state of the tree and collect statistics that are needed for performance and connection admission control. For instance, if congestion is building up at certain points on the tree, the SM may reconfigure the entire tree, deny access to new members, or lower the source transmission rate to alleviate the problem.

   The SM maintains two topology databases: Tree Topology which tracks group membership and tree topology information, and Domain Topology which tracks node and link state information for the entire domain. Each SM maintains its own Tree Topology image since they each manages its own multicast session. On the other hand, changes in the Domain Topology image are periodically downloaded from the DM. The update frequency is chosen such that it ensures that the local copy closely reflects the true network state while avoiding the need for excessive updates from the DM.

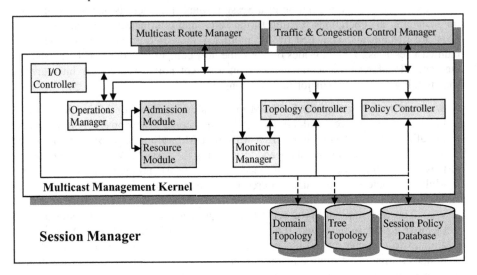

Fig. 3. Session Manager Architecture

Dedicating one SM per session naturally distributes the total computation load among many management entities, while still allowing multicast routing and other session-related decisions to be decided centrally which results in creation of more optimal multicast trees. Multicast connections are more complex and computationally intensive to manage than unicast connections.   Other than node and link state information, multicast routing and resource control also require tree structure and group membership information. Thus, to maximize the scalability of MRMA, it is important for the management architecture to maximize the distribution of the aggregate computational load across the network and to minimize the signaling overhead (i.e., eliminate the need for flooding). On the other hand, accuracy of the routing decisions is also important since it determines the efficiency in which network resources are utilized. A connection that is granted more bandwidth than its statistical mean requirement implies there will be less resources available for others, resulting in a non-optimum call-blocking rate. The accuracy of management decisions is dependent on the accuracy and completeness of the knowledge that the manager has of the true state of the network.

A fully distributed architecture may be effective in spreading the computational load across the network, but it is difficult for any one entity to obtain accurate and timely knowledge of the entire network, and so it suffers from the inability to make accurate management decisions. A central architecture, on the other hand, generally has complete knowledge of the entire topology and is therefore able to make more optimum routing and management decisions, but it does have the disadvantage of being a single point of congestion. The adverse effect is congestion on links located near the central manager. Moreover, as all network functions are executed at the manager, response times will experience long latencies as requests face long queuing times. MRMA's hierarchical arrangement of having global performance statistics monitored centrally by the DM, while session-specific statistics, which are of no use to other multicast groups, are handled by the corresponding SMs is a good comprise between the above two options. It offers a good balance between accuracy and scalability.

The physical location of each SM also greatly affects the load distribution. Placing all the SMs at a centralized location is definitely ineffective since it simply behaves like a centralized system. For source-based multicast trees, two possible options are: SMs reside on the access routers situated around the network perimeter, or SMs reside outside the network on the actual host/source computer. The first option achieves relatively good network load distribution, but it may still suffer from congestion for the case where an access router is serving many host computers that themselves are roots for different multicast trees. As such, the access router may be managing too many multicast sessions simultaneously and become a potential congestion point. MRMA implements the second option. Deploying the SM at the host computer maximizes the load distribution. In fact, it eliminates the task of setting-up and configuring the multicast session from the network altogether.

Element Agent. There is one Element Agent deployed on each network node. One of EA's primary functions is to manage the allocation of local node and link resources such as bandwidth, including handling resource reservation requests from the SMs. To guarantee Quality-of-Service, each EA tracks all the multicast trees that it is a part of and the amount of node and link resources that are reserved for each multicast flow. With that information, it can then apply the appropriate scheduling service for each flow. The EA's other major role is to monitor local node and link state information and periodically forward the summarized data to the DM and SMs.

Implementing resource management at the node-level evenly distributes the load of managing the network's resources. Moreover, by locating the resource manager close to the actual resource, the EA has all the necessary information needed to make timely and accurate resource allocation, routing and scheduling decisions locally.

3.3 MRMA Multicast Protocol

We also developed a QoS-sensitive multicast protocol to accompany the MRMA architecture. The Native MRMA Management Protocol (NMMP) specifies the execution procedures for all the MRMA operations. Note that although the current MRMA version supports NMMP, the general MRMA architecture can be readily customized to support other similar protocols such as QoSMIC [6].

[5] provides a detailed description of the protocol, but we shall introduce it here through an example application (see Fig. 4). We will describe how to multicast layered video over a MRMA-managed network. Layered video multicast protocols such as Receiver-Driven Layered Multicast (RLM) and Layered Video Multicast with Retransmissions (LVMR) generally have a source that transmits video as multiple

multicast layers/streams. Each receiver can than decide on the number of layers they would like the network to deliver to them, depending on its processing capacity. The more layers a receiver receives, the better the picture quality. For our example, the Src is the multicast source and R1...Rx are the receivers. The video is encoded into four layers L1...L4. Depending on the QoS desired, the receivers request the appropriate number of layers to receive.

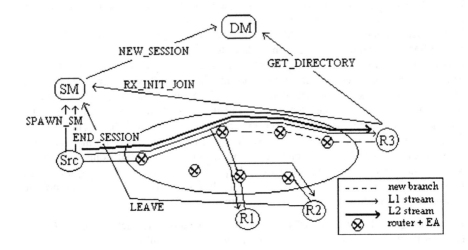

Fig. 4. Multicasting Layered Video over an MRMA-Managed Network

Create Session. Before multicasting can begin Src must spawn a session manager, SM, to initialize and register the session. Src also informs SM of desired session-specific QoS parameters such as maximum/minimum source rate (i.e., L1...L4 transmission rates), maximum number of group members, session duration, and etc. SM sends a NEW_SESSION message to DM to negotiate the QoS parameters on behalf of Src. DM makes the session admission decision based on the current network state (i.e., congestion level and available network resource level), and domain policies. If admitted, DM invokes the Address Manager to assign an unique multicast address to the session. DM then registers the new session with DM's Directory Manager so that its presence can be advertised to the entire network. The multicast session is now operational. Next, SM automatically invites any receivers that the source is interested in initiating multicasting with, if any. Otherwise, SM waits for new receivers to join.

Add New Members. We now assume the group already has two members, R1 and R2, and Src is multicasting layer L1 video to both. To join the group, the new receiver R3 first queries DM's Directory database for various session properties, including the multicast address and SM's IP address. R3 then sends a RX_INIT_JOIN message to SM containing receiver-specified QoS parameters (i.e., R3 wants to receive L1 and L2 layers).

Adding a new member requires SM to conduct admission-control and connection setup for a new tree branch. A new tree branch joining the new member to the tree would be part of the path, Path(Src, R3). Upon receiving the RX_INIT_JOIN message, SM first verifies if the receiver-specified QoS metrics satisfy session policies (i.e., there may be a policy that restricts non-local-domain receivers from joining multicast sessions that have a local-domain scope).

Next, the SM invokes the Multicast Route Manager to compute a feasible path connecting the existing tree to R3. The computation is based on state information from the Domain Topology and Tree Topology images, and the constraints defined by the receiver-specified QoS metrics. The selected route must have enough free bandwidth for transmitting the L1 and L2 streams. Once a feasible path is identified, SM attempts to reserve appropriate resources from all on-path nodes from Path(Src, R3). SM uses source-routing to send a RESERVE mobile agent that traverses the path from the Src to R3. The mobile agent can be a simple control packet or a control script. For each node it traverses, the mobile agent attempts to ensure the required resources are reserved and the multicast routing table is updated to reflect the new receiver. The affected nodes will include existing on-tree nodes and nodes comprising the new branch. If for some reason the state of any of the on-path nodes change during this process and it fails to reserve the required resources at a particular node, the SM executes a 'crank-back' mechanism [5] to release already reserved resources along the path. It can then invoke the Multicast Route Manager to re-compute an alternate route based on the new state information. Once path reservation succeeds, the join request is accepted and R3 can begin receiving multicast traffic. However, the join request is rejected if any of the above steps fail. Src now can begin multicasting both L1 and L2 streams, but only R3 will receive both streams.

The Source-initiated Join operation is very similar to the Receiver-initiated Join operation except SM must first invite a receiver to join.

Remove Members. With reference to Fig. 4, R2 wants to leave the group and so it must inform SM with a LEAVE message. Upon receiving the request SM queries its Tree Topology database to identify the tree branch that connects R2. To prune the branch, SM reverses the branch grafting operation. SM uses source-routing to transmit a RECLAIM mobile agent that will traverse along the tree from the source to the departing receiver. Upon reaching R2, the agent then propagates upstream along the reverse-path back towards SM. Along the way, it will deallocate the appropriate resources and remove the entry for R2 from the multicast routing table of each node it traverses. By reclaiming resources in a 'backwards' order fashion towards the source, we avoid the danger of pruning a branch too prematurely i.e. removing a branch entry from a upstream node's routing table before having the chance to successfully reclaim resources from all the downstream nodes.

Terminate Session. A multicast session can be terminated either by DM or the source. Upon receiving an END_SESSION message, SM terminates the session by removing all the group members, deallocating from the on-tree nodes all the resources that were reserved for the multicast session, removing the session entry from their multicast routing tables, unregistering the session from the DM's Session Directory, and finally killing the SM process. Much of this procedure is again conducted by multicasting a RECLAIM mobile agent through the tree.

4   Conclusions

In this paper we introduced MRMA, a hierarchical multicast management architecture that is designed to provision multicast-with-QoS service over IP networks.   Our architecture integrates all the key multicast functions (route selection, address allocation, session advertisement, congestion and transport control, tree monitoring and tree construction) into a single integrated management framework. It features a centralized monitoring mechanism that allows accurate gathering of domain topology information. It also introduces the concept of the Session Manager which allows multicast sessions to be managed on a per-session basis and distribute the aggregate computational load outside the network and onto the hosts.   Together the two mechanisms result in good management accuracy and scalability. Furthermore, the architecture models each of the above multicast components as stand-alone functional blocks that embed into a kernel core. The model allows the flexibility to 'swap-in' and 'swap-out' different implementations for each of the functional blocks, which facilitates leveraging of existing solutions.

For the next phase of our work, we plan to implement a MRMA prototype that can provision layered video multicast applications. In the near future, multimedia applications that require distribution of video over the Internet, such as video conferencing and video-on-demand, are good 'killer application' candidates, and multicasting the video in multiple layered streams is an effective technique for improving the fairness among the receivers. We hope to evaluate MRMA's support mechanisms for constructing multicast trees, for reserving resources and establishing QoS bounds on end-to-end delay, jitter variance, and data loss, and for scaling to accommodate growth. Overall, we hope to evaluate the architecture's overall ability to provision multicast-with-QoS service.

References

[1] S. Deering and D. Cheriton, "Multicast routing in datagram internetworks and extended LANs," ACM Transactions on Computer Systems, pp.85-111, May 1990.
[2] H. Eriksson, "The Multicast Backbone," Commun. ACM, vol. 8, 1994, pp. 54-60.
[3] S. Chen and K. Nahrstedt, "An Overview of Quality of Service Routing for Next-Generation High-Speed Networks: Problems and Solutions," IEEE Network, Nov/Dec. 1998.
[4] H. Holbrook and D. Cheriton, "IP Multicast Channels: EXPRESS Support for Large-Scale Single-Source Applications,: ACM SIGCOMM, Cambridge, MA, Aug. 1999.
[5] Ryan Wu, Irene Katzela, "Multicast Resource Management Architecture", University of Toronto, 2000.
[6] A. Banerjea, R. Pankaj, M. Foloutsos, "QoSMIC: Quality of Service sensitive Multicast Internet protoCol", SIGCOMM'98, 1998.

Operational Data Analysis: Improved Predictions Using Multi-computer Pattern Detection

Ricardo Vilalta, Chid Apte, and Sholom Weiss

IBM T.J. Watson Research Center
30 Saw Mill River Rd., Hawthorne N.Y., 10592 USA
Ph. (914) 784-7784
http://www.research.ibm.com/people/v/vilalta
vilalta@us.ibm.com

Abstract. Operational Data Analysis (ODA) automatically 1) monitors the performance of a computer through time, 2) stores such information in a data repository, 3) applies data-mining techniques, and 4) generates results. We describe a system implementing the four steps in ODA, focusing our attention on the data-mining step where our goal is to predict the value of a performance parameter (e.g., response time, cpu utilization, memory utilization) in the future. Our approach to the prediction problem extracts patterns from a database containing information from thousands of historical records and across computers. We show empirically how a multivariate linear regression model applied on all available records outperforms 1) a linear univariate model per machine, 2) a linear multivariate model per machine, and 3) a decision tree for regression across all machines. We conclude that global patterns relating characteristics across different computer models exist and can be extracted to improve the accuracy in predicting future performance behavior.

1 Introduction

Knowledge about the performance of a computer through time requires continuous monitoring. If the computer reports problems, information captured during this monitoring period can be of valuable help to determine the nature of the problem. Analyzing performance data through time may also help to decide how to fine-tune or refine the computer under analysis, which may ultimately lead to improved performance.

The automated process of monitoring the performance of a computer, storing all recorded data, analyzing the data, and generating results, is what in this paper we refer to as Operational Data Analysis (ODA). Such process is comprised by an area relatively unexplored: although mechanisms exist for monitoring a device or mechanism through time, little has been done to automate the process of extracting patterns from such data to predict potential failures, or to suggest an action plan that can guarantee optimal performance.

A. Ambler, S.B. Calo, and G. Kar (Eds.): DSOM 2000, LNCS 1960, pp. 37–46, 2000.
© Springer-Verlag Berlin Heidelberg 2000

This paper begins with a brief overview of Operational Data Analysis, a multi-step process highly relevant to systems management. The theme of the paper, however, focuses on a novel data mining approach for the time-series forecasting problem that extract patterns from historical data. The data-mining technique builds a multivariate linear regression model to the data using information extracted from a central database repository containing information about thousands of computers. We show how this model outperforms 1) a linear univariate model per machine, 2) a linear multivariate model per machine, and 3) a decision tree for regression across all machines. Our results provide evidence to believe that global patterns relating characteristics across different computer models exist. Taking advantage of such global patterns can improve the accuracy in predicting future performance behavior.

The organization of the paper follows. Section 2 explains each of the four steps comprised by the term Operational Data Analysis. Section 3 focuses on the data-mining step, describing a set of experiments that compare several models for the accurate prediction of machine performance. Section 4 shows our empirical results. Section 5 reviews related work in the literature. Finally, Section 6 presents our conclusions and future work.

2 Operational Data Analysis: Overview

Relevant to systems management is the automated process of predicting computer performance. One way to automate such process can be described in four basic steps (Figure 1). We describe the characteristics of each step next.

1. **Data Collection**. A mechanism in charge of taking snapshots reflecting the status of the computer must exhibit the following characteristics: a) snapshots should be taken every fixed interval of time, to ensure data consistency, b) the length of each interval should allow us to capture all important events. The length of each interval should provide enough granularity in the data, such that no important event is missed by the probes. Data collected by the probes may be averaged or aggregated before sending it to a data repository for analysis.

2. **Data Storage**. Information reflecting the status of a device through time is expected to yield large number of records. Furthermore, we expect to have applications where many devices are being monitored simultaneously. Consequently, data collected regularly over time should be stored in a common data warehouse. By consolidating all information in one place, data-analysis techniques can not only find trends on a single customer machine, but also search for correlations among different machines. This may lead to detect patterns that span across different models or even platforms (Section 3).

3. **Data Analysis**. Next is the application of data analysis techniques (e.g. machine learning, data mining, non-parametric statistics) to operational data. The idea is to search for patterns hidden in the data that may be helpful to understand how to improve machine performance. Data about the

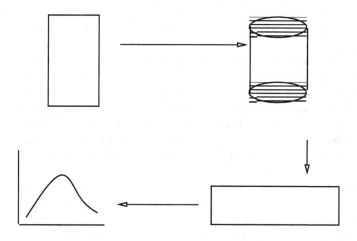

Fig. 1. The Four Steps in Operational Data Analysis.

performance of a machine through time may reveal that, at some point in the future, certain components may experience failure. This information can be important to avoid reaching conditions where it is known that certain component(s) will not work properly. The person in charge of the device or process may then be in good position to plan for possible upgrade scenarios.

4. **Generation of Results**. We may go beyond diagrams displaying statistics about the data, to include: a) the construction of a database enabling us to see results from different perspectives and under different search criteria, b) promotions and announcements; if the result of the analysis suggests the need for a machine upgrade, an opportunity exists to promote products that can satisfy the upgrade requirements.

2.1   System Description

The 4-step process described above has been successfully implemented in the automatic notification of computer performance to users of the IBM AS/400 platform. The prospective user starts by registering trough a web-based application. Separately the PM/400 monitoring software begins capturing performance data on the user's (AS/400) machine.

A central database contains information on the performance of all AS/400 computers that have been registered for this service. Each record in the database reports the values of tens of performance parameters for a particular machine,

month of the year, and shift, e.g., the average response time for machine with
serial number X, was on average 1.2 seconds in the month of July, 1999, during the morning shift. We limit ourselves to only twenty relevant performance
parameters as input for the analysis; the parameters are listed as follows: response time, maximum response time, cpu utilization, memory utilization, disk
utilization, disk arm utilization, system utilization, shift, avg. trans./hr, max.
trans./hr, avg. jobs /day, max. jobs /day, avg. IO Mb/sec, max. IO Mb/sec,
avg. interactive jobs, max. interactive jobs, avg. batch jobs, max. batch jobs,
avg. total transactions, max. total transactions. From the parameters above we
try to form predictions for the first six parameters: response time, maximum response time, cpu utilization, memory utilization, disk utilization, and disk arm
utilization.

When a user logs into the system, the four steps of ODA are performed.
Assuming enough information has been collected by the PM/400 monitoring
software, a remote connection brings data to a server where the data-mining step
results in a graph displaying (for each parameter) past performance, together
with predictions on future performance. Figure 2 shows an example of a result
graph.

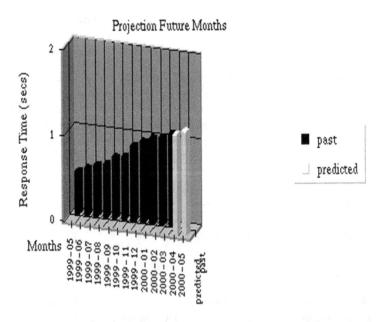

Fig. 2. The Resulting Graph after the four Steps in Operational Data Analysis
Have Been Performed.

3 Time-Series Analysis: A Data Mining Approach

The most challenging step in ODA is choosing the right data-mining technique for the application at hand. This section describes a series of experiments that compare several models for the time-series prediction problem. We wish to find a good model for the accurate prediction of future performance behavior. Our central claim is that an advantage is gained by extracting information related to multiple machines. Such global analysis enables us to extract patterns that help improve the accuracy of our predictions. In contrast, a model that is limited to information about a single machine is shown to perform less effectively.

3.1 Experimental Methodology

Our experiments use different models for prediction, e.g., univariate linear regression, multivariate linear regression, decision tree. These models attempt to capture the general trend of a dependent variable through time. We ignore periodic trends (seasonal analysis) or cyclic behavior, which we plan to address in the next project phase.

Some models were applied to data corresponding to a particular machine, while other models were applied to all the database at once. In the first case, we report average performance over all machines (Figure 3, Algorithm 1). If a particular machine has n records in the database, we take $n - 1$ records for training and use the last record for testing. In other words, we test the ability of the model to predict one month in the future. In the second case we transform the database by appending to each record the value of the target performance parameter for the next month (Figure 3, Algorithm 2). Assume a particular machine has n records in the database and that response-time acts as the target function. Record 1 for this machine is appended an extra target field corresponding to response-time in Record 2 (in the next month). This field becomes the target-function value to learn. The last record is evidently eliminated. We repeat this transformation process for all records in the database. We then train a model with $\frac{2}{3}$ of the transformed database, and test with the remaining $\frac{1}{3}$. We do this for each of the six parameters of interest (listed above), resulting in six transformed databases.

For each target parameter we report on the relative error. The relative error compares the error of the model with the error obtained by using the mean of all past values as our prediction. Relative error is the fraction resulting by dividing model-error by mean-error. We wish to see values much less than 1. Values above 1 implies we are better off simply using the mean value as our prediction.

In all our experiments we used the Data-Miner Software Kit [11]. Tests were done on an IBM RISC/6000 model 7043.

4 Results

Looking One Month in The Past. Our first experiment uses a very simple model as baseline for comparison. We take the value of the previous month to

Algorithm 1: Single-Machine Analysis
Input: Dataset D, size of dataset m,
Model M
Output: Average Accuracy A
SINGLE_MACHINE_ANALYSIS
(1)   Let $A = 0$
(2)   **while** (exist records in D)
(3)   Train M in $n - 1$ records
(4)   of same machine
(5)   Test M in last record n
(6)   Let $A = A+$ accuracy of M
(7)   **end while**
(8)   Let $A = A/$ no. of machines
(9)   **return** A

Algorithm 2: Multiple-Machine Analysis
Input: Dataset D, Model M
Output: Accuracy A
MULTIPLE_MACHINE_ANALYSIS
(1)   Transform database D by
(2)   appending to each record
(3)   the target variable one
(4)   month ahead.
(5)   Let the new database be D'
(6)   Let the size of D' be p
(7)   **while** (exist records in D')
(8)   Train M in $\frac{2}{3} \times p$ records
(9)   Test M in the other
(10) $\frac{1}{3} \times p$ records
(11)   **end while**
(12)   **return** Accuracy A of M

Fig. 3. Algorithm 1 Describes a Single-Machine Analysis. Algorithm 2 Describes a Multiple-Machine Analysis.

predict the value for the next month. We do this for each of the six relevant performance parameters. Table 1, Row 2, shows the results. Surprisingly this model is reasonably accurate, which we take to mean there is on average little variation in performance activity between consecutive months. A parameter with very low relative error is disk-arm utilization.

Single-Machine Linear Regression Model. The next experiment tests a single univariate linear regression model. As explained before, all records corresponding to the same machine, except the last, are used to train the model, the last record is used for testing. We then averaged over all machines. Table 1, Row 3, shows the results. Relative error is in all cases above one, and in three out of six cases above 2. Model performance is poor.

Table 1. Results for Relative Error.

Model	Relative Error					
	Response Time	Max. Response Time	CPU Util.	Memory Util.	Disk Space	Disk Arm
Previous Month	0.25	0.60	0.43	0.104	0.109	0.06
Linear Regression Univariate Per Machine	2.04	2.10	1.99	1.96	1.74	2.09
Linear Regression Mulivariate Per Machine	1.66	2.07	1.16	1.22	0.96	1.13
Linear Regression Multivariate All Records	0.27	0.49	0.28	0.00	0.00	0.00
Decision Tree All Records	0.54	0.54	0.26	0.01	0.00	0.07

Single-Machine Multivariate Linear Regression. Next we test a multivariate linear regression model per customer. To proceed we use the transformed database in which the value of the target function is appended at the end of each record. After averaging over all computers, we obtain the results shown in Table 1, Row 4. Most relative errors are above one. Although this model outperforms the univariate linear regression model per computer, accuracy remains below expectations.

All-Machines Multivariate Linear Regression. We now describe a series of experiments that employ all records in the transformed database. The general approach is depicted in Figure 4 (explained in Figure 3, Algorithm 2). The database is transformed by adding to each record the value of the target performance parameter one month ahead in time. The training set contains data for multiple parameters and thousands of computers. Applying a learning algorithm to this data is expected to capture patterns that span across multiple AS/400 computers.

The first experiment applies a multivariate linear regression model to $\frac{2}{3}$ of all available records. We used the remaining $\frac{1}{3}$ for testing. It is important to emphasize the main goal behind this analysis. The final model is intended to capture patterns not only relevant to each machine, but across machines. Table 1, Row 5, shows the results. Relative error is in all cases below one and for memory utilization, disk-space utilization, and disk-arm utilization, the error is zero with two significant figures. We notice also that most parameters have low variance which lead us to conclude that the multivariate linear regression model across all machines is a good model to predict performance behavior.

A similar experiment like the one above was conducted to predict 2-6 months in the future (tables not shown). In this case for most parameters relative error increases when compared to a one-month prediction, but remains stable after four months, and the relative error rarely goes above one. We conclude that even

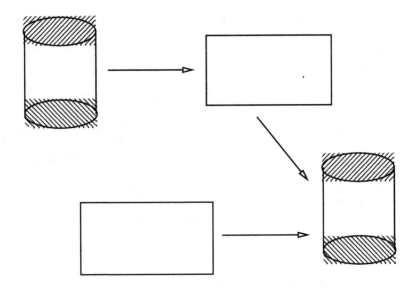

Fig. 4. The Process to Extract Global Patterns from the Performance Database.

for 2-6 month predictions, the multivariate linear regression method remains a good model of choice.

All-Machines Decision-Tree for Regression. Lastly, we tried a decision tree for regression over all records in the transformed databases. We used again the $\frac{2}{3}$ vs. $\frac{1}{3}$ split for training and testing. Results are shown in Table 1, Row 6. Overall the relative error for the decision tree is higher than the multivariate linear regression model. The decision tree performs well on memory, disk space, and disk arm utilization. We conclude a decision-tree model for regression does not outperform the multi-variate linear regression model.

5   Related Work

Computer performance forecasting is a topic of intensive research. Work is reported in the prediction of network performance to support dynamic scheduling [2], and in the prediction of traffic network [3].

An example of a study that uses past historical data to predict computer performance is reported by Yeh and Patt [4], where the goal is to produce a branch predictor to improve the performance of a deep pipelined micro-architecture. Similar studies are reported in the literature [5,9,10]. Compared to our approach these studies focus on predicting at the instruction level, whereas we contemplate the idea of predicting system behavior (e.g., response time, cpu utilization, etc.).

Zhichen et al. [6] predict the execution time of programs on parallel architectures. The approach is also data based, but is limited to predicting execution time alone.

A common approach to performance prediction proceeds analytically, by relying on specific performance models. For example, Hsu and Kremer [8] study prediction models at the source code level which play an important role for compiler optimization, programming environments, and debugging tools. Our approach is not model-based but data-based, which enables us to apply data mining techniques. Other studies have tried to bridge the gap between a model-based approach versus a data-based approach [7].

Apte and Hong [1] use rule-based induction in the prediction of financial market behavior. Their approach is akin to ours in that the model (rule-based) is applied to an entire database containing information about many entities (large capital firms) to predict future behavior.

6 Conclusions and Future Work

This paper compares the predictive accuracy of several models within the time-series analysis framework. Each record in the database under analysis describes the monthly-average performance of an AS/400 computer; our goal is to produce accurate performance predictions. Our experiments compare a univariate linear regression model per machine, a multivariate linear regression model per machine, a multivariate linear regression model across all machines, and a decision tree for regression across all machines. The multivariate linear regression model across machines shows the best performance overall.

An interpretation of the results reported in Section 4 can be advanced by comparing the difference between the two kind of models applied to the performance database: one applied to each machine separately vs. one applied simultaneously to all machines. Experimental results give a preference to the latter. We then address the question: why is there an advantage by looking for patterns that span across multiple computers? To answer this question it is reasonable to suppose that for this particular domain, machines that belong to the same model will experience similar performance if the overall machine utilization is the same. Thus, looking for patterns across computers increases the evidential support for correlations between the input variables and the target variable. For example, assume cpu utilization grows linearly with the number of batch jobs. If we produce a hypothesis based on data corresponding to a single machine, we may be able to support this linear correlation, but only with a limited number of records. In contrast, applying the same model using data from all machines, enables us to increase our confidence to decide if the linear correlation holds or if it was merely apparent due to the low number of records in the single-machine analysis.

Overall, we believe this kind of global pattern analysis deserves special attention, particularly today in which the amount of information stored in many central databases provides stringent data with which to produce solid inferences.

Future work will apply a larger variety of models to the entire performance database, and will attempt to derive a deeper theoretical explanation of the advantages that come by analyzing data corresponding to multiple computers.

References

1. Apte Chid and Hong Se June: **Predicting Equity Returns from Securities Data**. In Advances in Knowledge Discovery and Data Mining. Ed. Fayyad, U.M. and Pratetsky-Shapiro and Smyth, P. and Uthurusamy, R. AAAI Press, (1996) 541–560.
2. R. Wolski: **Forecasting network performance to support dynamic scheduling using the network weather service**. In Proceedings of the High Performance Distributed Computing Conference (1997).
3. K. Claffy and G. Polyzos: **Tracking Long-term Growth of the NSFNET**. In Communications of the ACM (1994).
4. Yeh T-Y. and Patt, Y.: **Alternative implementation of Two-Level Adaptive Branch Prediction**. In Proceedings of the 19th International Symposium on Computer Architecture, Gold Coast. Australia (1992) 124–134.
5. Brad, Calder and Dirk, Grunwald and Joel, Emer: **A system level perspective on branch architecture performance**. In Proceedings of the 28th Annual IEEE/ACM International Symposium on Microarchitecture. Ann Arbor, MI (1995) 199-206.
6. Zhichen, Xu and Xiaodong, Zhang and Lin, Sun: **Semi-Empirical Multiprocessor Performance Predictions**. TR-96-05-01, University of Texas, San Antonio, High Performance Comp. and Software Lab (1996).
7. Mark E. Crovella and Thomas J. LeBlanc: **Parallel performance prediction using lost cycles analysis**. In Supercomputing 94 (1994).
8. C-H. Hsu and U. Kremer: **A framework for qualitative performance prediction**. In Department of Computer Science, Rutgers University. Technical Report LCSR-TR98-363 (1998).
9. Brad Calder and Dirk Grunwald: **Next cache line and set prediction**. In ACM (1995) 287–296.
10. N. P. Jouppi and P. Ranganathan: **The relative importance of memory latency, bandwidth, and branch limits to performance**. In The Workshop on Mixing Logic and DRAM: Chips that Compute and Remember (1997).
11. Weiss Sholom and Indurkhya Nitin: **Predictive Data Mining**. Morgan Kaufmann Publishers (1998).

Scalable Visualization of Event Data

David J. Taylor[1], Nagui Halim[2], Joseph L. Hellerstein[2], and Sheng Ma[2]

[1] Department of Computer Science, University of Waterloo
Waterloo, Ontario, Canada N2L 6H1
dtaylor@uwaterloo.ca
[2] IBM Research Division, P.O. Box 704
Yorktown Heights, New York 10598
{halim,hellers,shengma}@us.ibm.com

Abstract. Monitoring large distributed systems often results in massive quantities of data that must be analyzed in order to yield useful information about the system. This paper describes a task-oriented approach to exploratory analysis that scales to very large event sets and an architecture that supports this process. The process and architecture are motivated through an example of exploratory analysis for problem determination using data from a corporate intranet.

1 Introduction

Effectively understanding the behavior of large, complex distributed systems requires appropriate visualization techniques to discover patterns that indicate underlying problems and the causes of those problems. While visualization is effective for modest sized data sets [4], displaying detailed data scales poorly. This paper describes a process for scalable visualizations for problem-determination purposes and proposes architectural principles to support that process.

We begin by illustrating the problem at hand. Consider event data extracted from a modest-sized intranet, say on the order of 50,000 events [2,3]. Discovering patterns such as periodicities (e.g., due to intermittent monitoring) and cold-start patterns can be done using simple displays, such as a scatter plot of all data by host name versus time.

This scales poorly for at least two of reasons. First, as we increase the number of events, there is a higher probability that events will overlay one another in the scatter plot, thereby diminishing the value of the detailed view of data. Second, while it is feasible to maintain an in-memory database of 50,000 events, larger quantities of events—5,000,000 or more per week is quite possible—cannot be efficiently handled in memory, at least not in a naive way. A third issue is the dimensionality of the data. With richer data, we have difficulties with the visualizations themselves since using two-dimensional renderings of high-dimension data makes it difficult to see complex relationships.

There are two possible approaches for dealing with these problems of scale. One is to design a visualization tool that deals directly with very large volumes of data, presumably keeping data on disk and using sophisticated algorithms both

A. Ambler, S.B. Calo, and G. Kar (Eds.): DSOM 2000, LNCS 1960, pp. 47–58, 2000.

to extract needed information from that data and to build optimal displays given
the available resolution. The alternative is to observe that it is very unlikely
a user will require detailed information from the complete set of data at one
time, particularly given the fundamental difficulties of comprehending a complete
picture of a very large set of data. Thus, if the user fundamentally needs to
extract subsets of data in order to obtain insight, a two-level visualization system
provides a good solution to both difficulties. In such a system, small subsets of
the data can be extracted as required and processed by the "core" visualization
tool, rather than trying to deal with all the data at once.

The concept is similar to "drill down" [1], but does not match it exactly.
The usual concept in drilling down is that multiple, hierarchically structured
categorical attributes are used in searching for areas in which a numerical attribute has extreme values. The exploration contemplated here includes such
situations, but also includes others. In particular, categorical attributes such as
server-up/server-down could be the target of exploration and the total number
of occurrences of server-down events under a certain set of restrictions might not
be the only relevant property of those events.

In this paper, we first present a scenario for analysis of large-scale data that
suggests it is feasible to proceed in the manner just described. Then, we gener-
alize from the scenario to determine principles for performing such analyses and
for the architecture of a software tool to support them. Finally, we describe our
plans to implement this architecture as an extension to the Event Browser.

2   A Scenario

The following scenario is based on actual exploration of data collected from a
corporate intranet. Numerous probes have been installed in the network to check
the health of important server machines. Each probe periodically checks on a set
of server machines, both by "pinging" the server and by attempting a user-level
transaction. In most cases, a server is checked by several probes. Probes are
frequently close, geographically, to the servers they are checking, but probes are
also located at considerable distances when the server is used extensively from
distant parts of the network.

Each probe attempt generates an event record indicating whether the server
was successfully contacted and, if it was, what the user-level and ping response
times were. In total, roughly 5,000,000 such probe records are generated per
week, which are stored in a relational database structured into a data warehouse
using a star schema. Extracting useful information from such a large collection,
beyond simple summaries such as average availability per server, presents significant challenges.

Here, we consider using this data for problem determination: locating servers
that are down or inaccessible or that have unacceptably long response times,
and beginning to identify the underlying problems.

The scenario presented here is intended to indicate possible approaches to
extracting useful information from this event-data warehouse. The techniques

described contain novel aspects related to the use of a two-level storage structure, but are not included primarily as interesting in their own right. Rather, they demonstrate the feasibility of the approach and provide a necessary foundation for the process and architecture discussions in the two following sections.

The first task attempted is to identify servers that are frequently reported down or inaccessible. In this case, we begin by extracting all the records, for the one-week period, that indicate the server was not successfully contacted by the probe. Fortunately, this occurs in about one per cent of the cases and thus provides an event set of manageable size for in-memory analysis.

Obtaining a list of servers ordered by total number of "down" reports allows us to investigate as many of the servers near the top of the list as seems appropriate. Most servers are checked by multiple probes. Thus, we use a subset of the data to plot the probes reporting that a single server is unreachable. The results are displayed in Fig. 1.

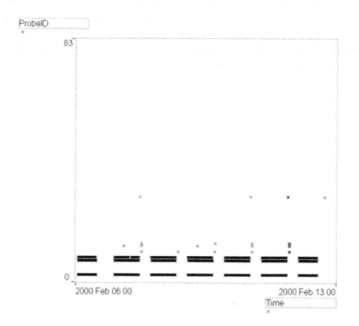

Fig. 1. "Down" Events for an Individual Server: Probe ID versus Time

Three of the probes appear to be reporting the server as down or unreachable nearly continuously and four others have reported it as down or unreachable on a few occasions. (Given the usual probe configuration, it is a reasonable assumption that the regular gaps represent periods that the probes are configured to be inactive.) A small "imperfection" in the second line for one of the three probes seems to indicate that it did succeed in contacting the server briefly. However, it is important to recall that we have simply extracted the "down" events, so the gap could represent "up" events or no events at all. In this case, extracting all the events describing probe attempts to that one server by that one probe reveals that it simply failed to report on the server at all for about two hours.

In this case, we can conclude with very limited use of event data beyond the "down" events that for most of the intranet there are occasional problems in contacting the server, but likely not frequent enough to be significant. However, either there is a significant network problem in the vicinity of the three probes (they are geographically close to each other) or they are misconfigured so that they are unable to contact the server. In either case, the situation warrants further investigation.

Now consider a second server. As shown in Fig. 2, there is a fairly small number of "down" events, reported by a large number of probes. These events appear to occur regularly at intervals of about one day. We confirm this by changing the x-axis to time-of-day, producing the display shown in Fig. 3. This confirms that the events are indeed clustered in a short daily interval; almost all of the "down" events occur between 8:30 and 9:30 A.M. A possibility is that some activity on either the server or the network occurs regularly at that time and triggers occasional failures. Since the time interval is early in the morning, it may also simply be a large amount of "getting started" activity as employees arrive for the start of the work day.

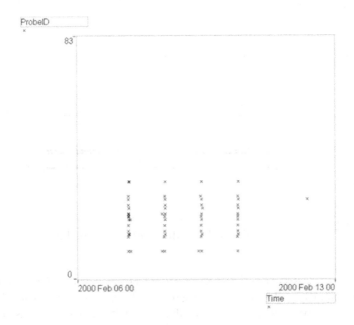

Fig. 2. "Down" Events for Another Server: Probe ID versus Time

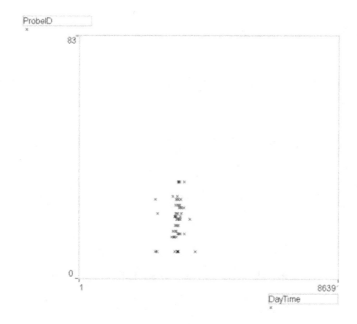

Fig. 3. "Down" Event for the Same Server: Probe ID versus Time-of-Day

A plausible theory for the time-out events is that there is an overload condition occurring regularly each morning. This overload then causes timeouts that produce the "down" reports. Pursuing this theory, it is appropriate to turn to the other aspect of our investigation and examine response times for the server in question. Doing so requires accessing "up" events for the server, so all events were extracted for that server which either reported it down or had above-average response times. Fig. 4 shows a plot of response time against time-of-day, for the "up" events in that set. Note that the x-axis, which in this figure and Fig. 3 gives seconds since midnight, here runs from 25216 (approximately 7:00 A.M.) to 79195 (approximately 10:00 P.M.), because it represents data for only one server and that server is only probed during that interval. In Fig. 3, the x-axis covers essentially the entire 24-hour day because data from all probes and servers is included. We observe that there is indeed a very high response-time peak; unfortunately it occurs from roughly 7:00 to 7:45 rather than coinciding with the "down" events.

The situation is thus rather mysterious, but clearly warrants further investigation. It appears that very early in the work day there is a burst of activity that pushes some response times into the several-minute range and that slightly later in the work day an unknown phenomenon causes occasional apparent server failures. (The reported failures are so brief that it appears unlikely a server could actually crash and restart, but the server does occasionally fail to respond to a probe.) To the extent that these represent an expected burst of activity at the start of the work day, they may simply need to be tolerated, but if any scheduled,

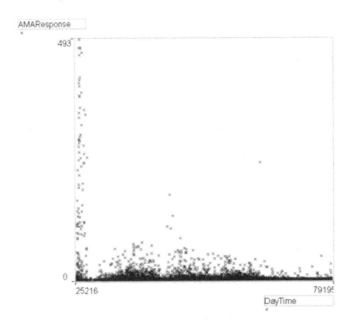

Fig. 4. Response Time versus time of Day, for the Same Server

non-interactive work is taking place during these periods, rescheduling it might alleviate the observed problems.

Finally, we can examine response times across the complete set of servers, rather than for this one server. For this purpose, events corresponding to response times three or more standard deviations above the mean were extracted, for all servers and probes. Fig. 5 shows response time versus (local) time-of-day for all probes and servers. In this case, the mean and standard deviation were computed for each probe/server pair, and response times were determined to be high relative to the statistics for that probe and server. Since some probes are located at the same site as the server and some are on other continents, averaging across probes, even for a single server, is not very useful in determining whether a response time is unreasonably high.

Some of the response times are quite extreme (almost 25 minutes), but the two most prominent features of the graph are a sparsely filled rectangular area and a prominent spike at the left side of the rectangle. The rectangle is easily explained as corresponding to the business day. It is hardly surprising that long response times tend to occur when most people are at work. The spike appears to be quite similar to that observed in Fig. 3. Here, the events in the spike have been colored so that they can be examined in other views, in particular to determine what servers they correspond to. In this case, the rectangular region selected as representing most of the spike (minus its base), contained 207 events, spread across 31 servers. However, somewhat closer examination reveals that 24

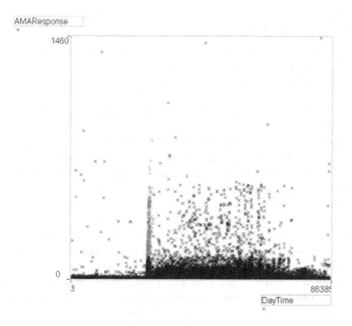

Fig. 5. High Response Times versus Time-of-Day, All Servers

of these servers, representing 196 of the events, are in the same country, but involve all three of the data centers in that country.

Thus, it appears that either the network or a substantial collection of servers in that one country is being subjected to some form of overload. A plot by time rather than time of day (not included here) indicates that the response-time peak is a five-day-a-week phenomenon. Thus, the problem with response times first observed for a particular server that generates an odd pattern of "down" reports turns out to be common to numerous servers, but only within a particular country.

This section has briefly described some of the approaches used in extracting useful information from the event-data warehouse for the corporate network. All of the work described was performed by formulating a SQL query, using it to extract events from the database, reformatting the data to make them suitable for use with the Event Browser, and then using the Event Browser on the data. Clearly, a more tightly integrated approach is highly desirable. The next two sections derive general principles from this scenario, first for the process to be followed in analyzing such large event sets and then for the architecture of a tool to support that analysis.

3   Process Principles

The scenario described in the preceding section suggests a general process that can be followed in dealing with large volumes of event data. The process can be briefly outlined as follows:

1. Apply any basic restriction (such as a date range) and obtain summary statistics.
2. Using the summary statistics together with domain information and the desired goal for exploration, formulate an initial set of restrictions and extract the corresponding subset of events. If all obvious initial restrictions yield sets that are too large, a further possibility is to extract a statistical sample of the specified events, for initial analysis.
3. Repeat steps 4 through 6 until analysis is complete.
4. Explore the current subset of events, using any appropriate interactive facilities available. Look for unusual (extreme) values, look for patterns in appropriate visualizations of the data, and so on.
5. If phenomena of interest are discovered but cannot be fully investigated in this event subset, determine (a) what common properties are possessed by the events of interest and (b) what restrictions used in extracting this subset need to be relaxed in order to obtain further information relevant to the phenomenon discovered.
6. Based on the analysis of the preceding step, formulate a new set of restrictions, extract the corresponding subset of events, and (a) replace the current set of events with the new set, (b) form the union of the current and new sets, or (c) keep both new and old sets for concurrent but independent examination.

Most of the above follows quite directly from the scenario described in the preceding section except for the first step and the three-way choice in the last step. In the scenario, the first step was effectively replaced by an understanding that "down" events and events with high response times are quite rare.

For the last step, choice (a) was used throughout, that is, a new set of events was identified and used in place of the previous set of events. A difficulty with that strategy is that a user will be trying to accumulate information about a problem until the problem is fully understood or all available data are exhausted. Simply discarding one event subset and replacing it by another can cause a loss of context and accumulated information.

The somewhat complicated specification for the events used to create Fig. 4 is essentially intended to maintain context. By extracting the "down" events previously examined as well as the events with long response times, the new events can be seen in the context of the old events. A user would not need to write such specifications, which could become progressively more intricate as analysis progressed, if it was possible to add new events to the existing set or to display two (or more) event sets simultaneously. The ability to display two sets simultaneously also facilitates taking a quick exploratory look at a new subset and then returning to the set currently being examined.

4   Architectural Principles

An effective tool for examining very large sets of event data must provide good support for the process outlined in the preceding section. The remainder of this section lists a number of principles that appear necessary or at least very desirable in designing a tool that will support the kind of analysis discussed.

Principle 1: Provide a general, easy-to-use mechanism for specifying event subsets, allowing all event attributes to be used in the specification.

Repeatedly selecting event subsets is a central feature of the process discussed in the preceding section. The scenario indicates that there can be a great deal of variation in the criteria used to specify a subset of events. At minimum, all the attributes of an event need to be useable in the selection criteria. For categorical attributes it may be sufficient to allow selection of the individual categorical values to be included, but see Principle 3 for categorical attributes with a very large number of values. For numeric attributes, in addition to ranges with constant limits, range end-points that are simple functions of properties like minimum, maximum, mean, and standard deviation are also needed. It is also necessary to provide a mechanism for specifying over what set the statistics are to be computed. In the scenario of Sect. 2, a specification would indicate that means and standard deviations were to be computed for each combination of probe, server, and up/down status (the last in order to exclude fictitious response times reported when no response was received). The mechanism for specifying new subsets should also be appropriately integrated with other mechanisms for examining event data, so that selections made in examining the current event subset can be easily used in selecting the next subset.

It is somewhat less clear what facilities need to be provided in combining restrictions on different attributes. The subsets used in the scenario were almost all specified as the intersection of restrictions on the relevant attributes. Only one involved a union: all the events for a particular week and a particular server that either reported the server down or had a high response time. The scenario does not provide a strong case for using a union since the "down" events were filtered out of the graph in Fig. 4, but a graph could have been presented in which the "down" events were marked with color and their position could then be directly compared with the position of the response-time peaks. If a general facility is provided for adding newly extracted events to the current set (as required by the principles of the preceding section), then a union operation need not be provided directly. This will make the use of union somewhat less convenient than using specifications based purely on intersection, but that may be a reasonable tradeoff given the likely frequency of use.

In most cases, the complete collection of event data will be in a database with a SQL interface, so it is also possible to allow a user to write a SQL query directly and use that for data extraction. While it would be undesirable to require ordinary users of an event-browsing facility to write SQL, such an escape mechanism could be useful to avoid overburdening the graphical user interface with seldom-needed capabilities.

Principle 2: Event-subset specifications should allow direct specification of the size of the result subset.

One issue not explicitly discussed in Sect. 2 was the varying definition of "high response time." In one case it is simply response times greater than the mean and in another response times three standard deviations above the mean. Not surprisingly, the underlying reason in each case is to provide a restriction that appropriately limits the size of the specified subset. In one case, only a single server is involved so a fairly loose interpretation of "high" yields a subset of reasonable size, but when extracting events without restriction by probe or server, a very stringent definition of "high" is required in order to limit the subset size. Some trial and error went into the determination of these specifications, which it would be desirable to eliminate. A facility that allowed a specification like "the events with the highest response times, which also satisfy this set of selection criteria" could be used to obtain a set of the appropriate size for interactive browsing. Unfortunately, this raises yet another difficulty: what does "highest" mean? It could be highest in actual value, highest as a multiple of the mean, or highest in number of standard deviations above the mean. Again, it is necessary to give the user considerable flexibility in specifying constraints because no one form will be suitable for all purposes.

Principle 3: The hierarchical structure of categorical attributes needs to be made readily accessible to the user.

Another issue is implied by observations in Sect. 2 like "they are geographically close to each other" and "24 of these servers ... are in the same country." In this case, stereotyped naming of most servers and all probes allows a knowledgeable user to deduce geographic information, with some difficulty, directly from the names. It is undesirable to ask users to assimilate such a scheme and in other cases, names will not be stereotyped, requiring access to on-line or off-line reference material to determine relationships. In this case, the event-data warehouse includes a six-level geographic hierarchy; similar hierarchical information is likely to be available whenever a categorical attribute has a very large number of values. Making such hierarchy information directly accessible when using an event browser would simplify the selection of events using categorical attributes with many values, as well as being helpful in determining the significance of patterns observed.

A facility for accessing such hierarchical information needs to be quite flexible. In particular, multiple hierarchies can reasonably exist for a single attribute. For example, in the case of the servers in the corporate intranet, in addition to the geographic hierarchy there are separate, shallower hierarchies corresponding to the function of the server (database, mail, web, etc.), hardware architecture and speed, and operating-system type and version.

Principle 4: Provide summary data, using an interface similar to that used for displaying detail data.

In many cases, some information about the overall set of events will be needed before it is possible to formulate a reasonable specification for the initial set of events. For example, in the analysis described in Sect. 2, a SQL query could have been used to determine the number of "up" and "down" events for each server. Then, rather than extracting all "down" events and working with them, the events for servers frequently reported down could have been examined for each server in turn. If the set of "down" events had been substantially larger for any reason, including the desire to examine a period longer than a week, such summary data would have been vital.

As much as possible, the usual interface for examining detailed data should be used with summary data, rather than providing a separate interface. Although only a subset of the usual functionality can be provided for such summary data, it will clearly be helpful to the user if it is not necessary to learn the use of a separate interface. Such summaries should be available for event subsets as well as for the entire set of stored events. For example, if all subsequent analysis is for events in a particular week, then summary information needs to be restricted to that week as well.

Principle 5: Multiple event subsets should be simultaneously accessible to the user.

The discussion in Sec. 3 indicates that simultaneous access to multiple event subsets can be very helpful. It might appear that providing such a facility would require that the total size of the event subsets be no larger than a single subset would be, to obtain good performance. Fortunately, this may not be true. In most cases, the central performance issue is not likely to be the total virtual-memory footprint but the use of virtual memory when scanning through an event set to perform operations like coloring all events (potentially in several displayed views) that have a specified property.

Principle 6: User actions in selecting event subsets should be recorded and made accessible.

As a user performs a lengthy sequence of selecting and examining event subsets, the current state of the present and previously examined event subsets may no longer be clear to the user. This is particularly likely if multiple subsets are retained for concurrent examination or subsets are modified by adding additional events after they are first created. To allow a user to understand the state behind the current displayed views and also to allow a user to retrace previous steps if necessary, a "history" facility is highly desirable.

5 Conclusions and Further Work

In this paper, we have determined both an effective process for working with large sets of event data and a set of architectural principles for building a tool that properly supports that process. With the exception of concurrently examining multiple event sets, the process described can be used with the current Event Browser, but the process is quite inconvenient. The chief problem is that extracting an event subset requires writing a SQL query, executing it, running a program to reformat the output of the query, then opening the resulting file in the Event Browser.

Our plan is to extend the Browser to allow the suggested process to be performed interactively during a single Browser session. Most of the work required is straightforward, for example, the SQL queries required can be readily generated from standard patterns, with user input essentially used to "fill the blanks" in those patterns. The major challenge is likely in devising a user interface that provides enough flexibility to implement all the principles described in the preceding section without overwhelming the user. Additional facilities not covered by the principles of Sect. 4 might also be added, for example a recording facility that could be used to document the actions taken or to create a "macro" that could be used by a less experienced user in later performing a similar analysis.

At least two additional challenging problems remain to be addressed. One is that this paper has only considered scaling issues with respect to the number of events available. Another potential problem is scaling in the number of attributes per event. An implicit assumption in the above is that the number of attributes per event is small enough that the user can individually inspect any that appear of potential interest without special support from a software tool. We have recently been given access to a new set of data in which the number of events is not particularly large but there are more than 250 attributes per event. Different techniques are required to help the user extract useful information from such a large set of attributes.

Another problem is the development and integration of techniques for mining the event data for patterns that are not immediately apparent in standard views such as scatter plots and bar charts. Some mining techniques have already been developed for use with the set of data currently being examined by the Event Browser, but further techniques are required and they also need to be made workable in a two-level environment as proposed here. That is, it should be possible to execute a mining algorithm against an entire event-data warehouse or a large subset of it, probably as an off-line activity because of the time required, and then interactively use the results of the mining together with the other "basic" information available for the events.

References

1. Robert F. Berry, Joseph L. Hellerstein: A flexible and scalable approach to navigating measurement data in performance management applications. Proceedings, Second International Conference on Systems Management, Toronto, Canada (June 19–21, 1996).
2. M. Derthick, J. A. Kolojejchick, S. F. Roth: An interactive visualization environment for data exploration. Proceedings of Knowledge Discovery in Databases (1997).
3. Sheng Ma, Joseph L. Hellerstein: EventBrowser: A flexible tool for scalable analysis of event data. Proceedings, Distributed Systems, Operations and Management, Zurich, Switzerland (October 11–13, 1999).
4. Edward R. Tufte: Envisioning Information. Graphics Press (1990).

Service Management Platform: The Next Step in Management Tools

Gabi Dreo Rodosek

Munich Network Management Team
Leibniz Supercomputing Center
Barer Str. 21, 80333 Munich
dreo@lrz.de

Abstract. IT service management is a new, upcoming area of research with several new challenges to cope with. The paper uses a systematic approach to identify new requirements to service management and analyzes some of them more precisely. As a result of the discussion, the introduction of a *service management platform* is motivated. The architecture of a service management platform is specified as well as the integration aspect discussed on a fault management scenario.

1 Introduction

A plethora of new challenges to management is arising due to the paradigm shift from network and systems management to service management. Trends of evolving electronic services, e-commerce, e-business, outsourcing, liberalization and open service market are the drivers of an IT service management. IT services are more and more treated as products with quality and price and it becomes more than ever important to achieve overall management solutions applied across all types of networks, computing systems and applications.

Network and systems management have a device-oriented view on the managed environment. In other words, network and systems management refers to the management of the infrastructure. Management information is specified in terms of MIB variables for a device (e.g. IfInOctets) or application. With the paradigm shift to service management, a new dimension is added. Managed objects are now not only network devices, end systems or applications but more complex objects, namely services. A *service* in our context is comprised of several other services, applications, devices and end systems. It is provided to a customer with a certain Quality of Service (QoS) as agreed upon in Service Level Agreements (SLAs). Thus, a service is provided to various customers with various QoS parameters. For example, a WWW service is provided to one customer with an availability of 99,7% and to another with an availability of 99,8%.

Managing a service means to deal with other services, several applications, devices and end systems as well as dependencies between them. One of the most challenging issues of service management is to recognize what are the service components and how a service is realized (i.e. provisioned) on an infrastructure.

A. Ambler, S.B. Calo, and G. Kar (Eds.): DSOM 2000, LNCS 1960, pp. 59–70, 2000.

In other words, the problem of mapping the service layer to the resource (e.g. network devices, end systems, applications) layer needs to be approached. Examples from the area of performance and fault management show the associated problems with this mapping more precisely. In the performance scenario the mapping of QoS parameters - as agreed in SLAs - to technical parameters and thresholds - as reported by management tools - is discussed. The fault management scenario deals with the questions how to aggregate events from network devices, end systems and applications in order to obtain the status of a service.

Service quality, external provisioning of services, supply chains are examples of other aspects of IT service management. A systematic top-down analysis and the identification of these new research questions are the first steps towards service management. As a result of our discussion, the need for a service management architecture and a service management platform is identified. The objective of the paper is to motivate the development of a service management platform.

Work in this area refers to topics such as the specification of a service architecture for telecommunication services, including some management aspects in the management architecture as specified by TINA [1], the specification of an information model [2], deals with service quality aspects (e.g. [3], [4]) or analyzes the problem area from a process-oriented view as described in Telecom Operations Map (TOM) [5]. Vendors have as well recognized the need to provide a service view on the device-oriented managed environment as demonstrated with Business Process Views from CA, Global Enterprise Manager from Tivoli or SLA reporting tools such as Network Health from Concord or InfoVista from InfoVista. However, existing approaches are only (rudimentary) add-ons to existing device-oriented network and systems management tools. They approach the problem of providing service views from a bottom-up side in terms of what service views can be represented based on the available device-oriented data. By approaching the problem area of IT service management from a top-down perspective, the need for a new management platform, the *service management platform*, becomes obvious.

The paper proceeds as follows: Section 2 identifies aspects of IT service management. Based on these requirements, some of the new challenges and research questions of service management are identified. A particular one, namely the architecture of a service management platform, is discussed in Section 3. In Section 4, the integration aspect of a service management platform is explained on a fault management scenario. Finally, in Section 5, some concluding remarks and interesting open issues of IT service management are sketched.

2   Aspects of IT Service Management

To identify the requirements of service management, we discuss the problem area on a scenario. Fig. 1 motivates and gives us a picture of the topics to discuss. As already stated, a service in our context is a functionality provided to a customer with an agreed QoS at the customer-provider interface. A service may be comprised of other services, applications, network devices and end systems. Services can be provided by external providers or internally within an organization. As depicted in Fig. 1, four types of aspects provide the basis for a systematic analysis:

- organizational aspects,
- service functionality aspects,
- service quality aspects, and
- management aspects.

Organizational aspects refer to several issues. Roles (customer, provider) are certainly one of them. An organization can act in a customer or provider role or in both roles simultaneously. In case an organization is fulfilling both roles simultaneously, it acts as a value-added service provider, using one or more services to set up the value-added service. A service can be provided (in whole or in parts) internally within an organization or externally by external providers. In the discussed scenario in Fig. 1 organization B provides two internal services (DNS and IP) and uses one external service (ATM) from organization C to set up the WWW service. Nevertheless, whether a service is provided internally within an organization - for example from another department - or externally by another provider, the service should be provided with an agreed QoS.

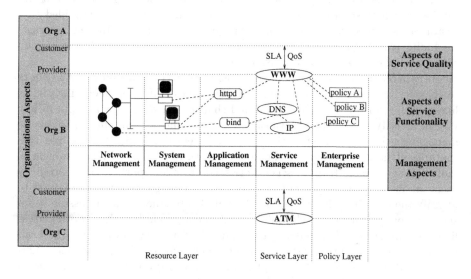

Fig. 1. Scenario

Customer-orientation is another important issue to consider. In our scenario, WWW services of organization B are provided to many customers. For each customer, individual SLAs and QoS parameters have to be specified and enforced. In case these Web services are set up on top of the same infrastructure (devices, servers etc.), a strict logical customer separation is necessary. In case a customer requests a strict physical separation, dedicated devices for each customer need to be provided.

Functional aspects refer to the

- functionality of a service and functionality of its components, such as other (sub)services, applications, end systems or network devices as well as
- dependencies between them.

An approach to describe the dependencies between services (e.g. WWW service depends on DNS and IP) is in terms of a service dependency graph as proposed in [6].

Quality aspects refer to service quality as agreed between a customer and a provider in a SLA. SLAs contain all service-related, organizational and operational information and parameters which are required to describe quality aspects of a service. Certainly, the specification of service-specific QoS parameters as well as the methodology for their measurement and verification are one of the most important ones. Open issues hereby are for example the building blocks of a SLA, SLA management as well as the management of the infrastructure to verify SLAs. One of the objectives of SLA management is to aggregate specific threshold values of devices in a dedicated way to obtain service-oriented parameters, like throughput or availability. Another topic is the specification of service catalogues where several services can be combined together and provided to a customer as one service. An example is the provision of an "Internet service" (including a Web service, DNS, proxy etc.). Classes of service quality (e.g. gold, silver) are other aspects to be consider as well.

As each object needs to be managed, there are several **management aspects** to consider for service management, too. The service lifecycle, including phases like planning, installation, operation and change of a service, is an appropriate way to identify these aspects. Another view to identify requirements to service management is with respect to the functional areas (FCAPS). An example from the area of fault management should demonstrate this. A typical problem of fault management is to map service-oriented trouble reports - as reported by users - to device-oriented events - as reported by management tools - in order to identify the cause of a fault. As already stated throughout our discussion so far such a mapping is not only relevant for fault management but also for other areas like performance or SLA management. The identified aspects provide the basis to systematically identify and refine requirements to service management. This is discussed in more detail for the planning and operation phase. A summary of the discussed aspects is shown in Fig. 2.

The planning phase is certainly one of the most interesting phases of the service lifecycle. Planning was always an issue, also in network and systems management. Although there was always a need for planning support to decide for example whether to expand an international link, change peerings, change the topology or expand the proxies, planning decisions were in most cases done based on experience and knowledge of network and systems managers. The reason is that the usage of planning tools (e.g. simulation tools such as [7]) is in most

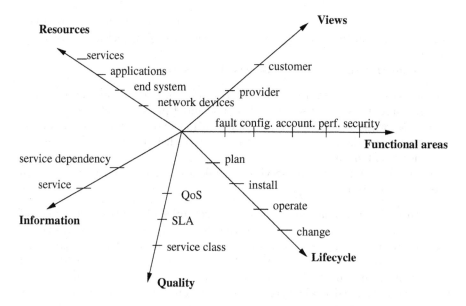

Fig. 2. Aspects of IT Service Management

cases associated with a lot of effort to collect all the necessary - sometimes even not available - data to obtain feasible results.

Nowadays, the importance of planning for the IP service has been rediscovered in terms of IP traffic engineering [8]. The objective of IP traffic engineering is to improve the perceived quality of network services and at the same time to maintain a high level of resource utilization. In spite of Gigabit speeds or even because of that dramatic increase in backbone speed, new precise control over traffic flows in terms of end-to-end management is essential. It should be noted that traffic flows represent *service-oriented* information of the IP service and are necessary to recognize the utilization of the backbone for the purpose of planning (e.g. [9]). What-if scenarios to experiment with configuration changes and traffic optimization are examples of planning issues.

Although planning was already a necessity for network and systems management, it is a must for service management. Adding a new service means to add new applications, network devices, end systems or change the existing resources to cope with the additional load. Examples of planning questions a manager is confronted with are as follows. Will the addition of a new service or new customers influence the quality of the already provided services? What are the potential consequences? What changes in the infrastructure are necessary (adding new network devices, enhance bandwidth, change topology, add new servers, change policies) in case new services with certain qualities need to be provided?

Due to the distributed realization of a service and the dependencies between its components and the resulting complexity, such planning decisions need to be tool supported.

To summarize, the planning phase has identified the following research topics:

– It is necessary to obtain *service-oriented information* for planning purposes (e.g. traffic flows for IP service). The associated problems refer to the questions (i) what is service relevant information and (ii) how to obtain the necessary raw data as well as how to aggregate them to obtain the necessary service-oriented information.
– Due to the complexity and service dependencies, the provision of a decision support tool is a necessity.

Another phase to analyze is *operation*. Examples of the research topics of this phase are the provision of a service view for the (i) service provider, and for the (ii) customer in terms of a Customer Service Management (CSM) [10]. The first point deals with the issue to provide a service view on the device-oriented management environment which requires to map the service layer to the resource layer. Some associated challenges to cope with are (i) to map service quality parameters to technical parameters and (ii) to aggregate device-oriented events to recognize the status of a service.

With the shift to service management and especially SLAs customers do request a view of their subscribed services as well. Customers want to have either scheduled reports about the quality of their services or even online access to the present quality information about their subscribed services. Besides, they want to report problems in a smooth and easy way. Such a customer view can be realized and provided to customers in terms of a CSM. CSM is a part of service management, and should be considered as a separate service.

Associated with the operation phase are policies. A service provider has to specify and enforce policies in order to describe constraints on the operation of services (e.g. some pages of a web server are only accessible from certain subnets). However, this is out of scope for our discussion.

From the discussion above, the following research topics can be identified:

– mapping the service layer to the resource layer (network devices, end systems, applications) and
– to provide (different) service views on the managed environment to providers and customers.

The analysis so far has identified requirements to service management with respect to the analyzed phases of the service lifecycle. In order to identify further research topics other identified aspects needs to be analyzed accordingly.

3   Architecture of a Service Management Platform

Although the previous discussion identified only some of the requirements to service management, the need for a service management architecture and as a consequence for a service management platform became obvious. The specification of a service management architecture is certainly a non-trivial task and is not the focal point of our discussion. The objective of this paper is to motivate the need for a service management platform in order to address the previously mentioned requirements and resulting research topics.

The architecture of a service management platform should be specified in analogy to the architecture of network and systems management platforms. With respect to the architecture of a management platform [11], a service management platform should have the following elements (Fig. 3):

- An infrastructure
  The infrastructure should consists of a repository where the service management information (i.e. the Service MIB) is stored and a standardized communication middleware such as CORBA. The term MIB should be understood in a more wider sense as a Management Information Base which contains complex objects such as services and not necessarily only SNMP MIB variables. A service MIB should describe the service functionality, the dependencies between the components of a service and the associated service-related QoS parameters.
- Basic applications such as:
  - a Service MIB browser,
  - an event management application as well as
  - a service level management application.
  A Service MIB browser provides the capability to load the description of a service into the platform. It is the same as loading a SNMP MIB into a network management platform.
  The objective of the event management basic application is to aggregate device-oriented events - events from the network and systems management platforms - in order to obtain and visualize the status of a service. A service could be for example in the status up, down or degraded in quality.
  The service level management application is concerned with the mapping of service quality parameters to technical parameters like thresholds of a resource. For example, the availability of a service is determined by an appropriate aggregation of technical parameters and thresholds of service components (i.e. resources).

Beside the infrastructure and the identified basic applications, several so-called management applications can be identified from the previous discussion. It should be noted that the reason if an application is considered to be a basic application is whether it is used by other management application. For example, event management provides information for the correlation of user trouble reports or provides the necessary information for the CSM GUI and SM GUI, as depicted in Fig. 3.

Service planning is one of the management applications. It resides on top of a service management platform and enables e.g. what-if scenarios by using for example constraint-based approaches to support planning decisions. Potential

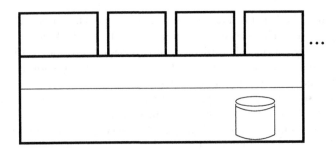

Fig. 3. Architecture of a Service Management Platform

problems and consequences could be identified in case new services and/or customers are added or changes on the service layer could request changes of the underlying infrastructure. Another planning issue is to "move" services around the infrastructure. In other words, it is necessary to support migration of services for example from one server to another. Of course, services are complex objects residing on several resources, thus it is necessary to consider that the migration involves several resources. Such a feature is associated with several issues from trivial ones (e.g. enough space on the target server) till complex planning issues and configuration changes. Constraint-based approaches are certain a reasonable way to address this problem area (e.g. [12]).

As already mentioned, a service management platform should provide **service views** on the managed environment to the provider itself and to the customers. Thus, on top of a service management platform (Fig. 3) the following GUIs need to be provided:

- the GUI for a service provider (SM GUI),
- the GUI for the customers (CSM GUI).

CSM as described in [10] realizes a lot of functionality which in fact belongs in a service management platform. The aggregation of device-oriented thresholds in order to obtain service-oriented parameters for the IP service is certainly an example of a basic application, namely the service level management. Because a service management platform is missing so far, the mentioned functionality has been realized as a part of the CSM application. As soon as a service management platform is developed, the current CSM application and implementation is reduced to a simple CSM GUI (as shown in Fig. 3), providing *one* interface to the customer. Customers use CSM to (i) inform themselves about the quality of their subscribed services, and (ii) can actively report problems over one interface. Thus, Web interfaces to trouble ticket systems or the Intelligent Assistant [13] should be part of CSM.

With respect to the discussion so far, the architecture of a service management platform can be specified as shown in Fig. 3. Examples of basic applications of a platform are event management, service level management and a

Service MIB browser. Examples of management applications are service planning, a CSM GUI and a SM GUI. The CSM GUI refers to the customer view of a service whereas the SM GUI refers to the service provider view.

Another important aspect of a service management platform are development tools, such as an editor for the generation of new service descriptions, descriptions of QoS parameters and the generation/updates of the service MIB.

To summarize the previous discussion, the functionality of a service management platform should provide:

- the ability to deal with service descriptions in terms of a Service MIB (i.e. it is necessary to load/unload the service MIB in the service management platform),
- the ability to describe services and dependencies between service components (i.e. development tools like editors, compilers),
- to provide service views on the managed environment to the provider and customers, and to
- map services to resources (e.g. to aggregate events from network and systems management to service-oriented events, to map QoS parameters to technical parameters of devices).

Throughout the discussion the central point was always the mapping from the service layer to the resource layer. Of course, such mapping needs to be done automatically and tool supported. A precondition for this is to describe characteristics of resources in an appropriate way (e.g. server with Gigabit Ethernet interface with several CPUs). Besides, the service requirements to resources need to be specified as well (e.g. a SAP service for a large enterprise with an availability of 99,9%). To approach this problem means to specify a framework for dynamic service provisioning. This is, however, out of scope for this paper.

The service management platform has an important role to play in the management environment. It has the role of a **master** for the network and systems management platforms. This means that the service management platform sets for example thresholds of devices, event configurations of the network and systems management platforms. In fact, it configures both device-oriented platforms. Therefore, it is necessary to standardize ("smart") interfaces to both device-oriented managers. The following section should demonstrate the integration of the service management platform with device-oriented managers on a correlation scenario.

4   Applicability of the Service Management Platform

With the introduction of a service management platform the integration issue becomes even more important than in today's management. An impression of the new challenges should be given by the following example from the area of fault management.

The necessity to map service-oriented views to device-oriented is a hot topic in today's management, especially in fault management. Users report their problems from a service point of view. They report that they can not send an email, have problems to access a web site or can not send orders in SAP R/3. On the other side, network and systems manager have a device-oriented view of the managed environment in terms of ups and downs of nodes, links or running processes on dedicated workstations. The objective of fault diagnosis is to map both views in order to recognize the cause of the fault.

To be able to map these views, several problems need to be approached. From the device-oriented view, the problem is the enormous amount of events to deal with as reported from management tools. To cope with the enormous amount of events, event correlation tools like InCharge from SMARTS or Tivoli Enterprise Console (TEC) are used to correlate events. On the other side, users report problems in most cases in an unprecise manner (an approach to deal with this problem has been presented in [13]).

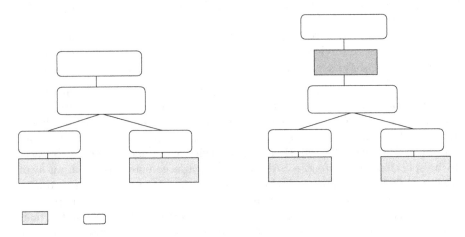

Fig. 4. Levels of Event Correlation

Event correlation is performed on various levels (as depicted in Fig. 4 (a)) with respect to topological, temporal and functional criteria. Mostly, the correlation of network events is performed as part of a network management platform (e.g. a down/up trap of a node within a certain time interval should be suppressed). Such correlation tools are mostly a part of a management platform (e.g. ECS from Seagate as a part of HP OpenView Network Node Manager and IT/O, InCharge from SMARTS as a part of Tivoli NetView). Another level is the correlation of system events (e.g. wrong passwords). The next level is the correlation of network and system events with respect to topology. For example, a reason that a workstation does not respond to a ping could be that the workstation is not working correctly or that there is a problem with the connectivity. To recognize the cause of the fault, network and system events need to be correlated and sometimes actively polled in addition.

The last level of correlation is done with respect to functional criteria. The knowledge about the dependencies between services is nowadays hardcoded for example in rules of TEC, based on the experience and knowledge of some experts. The generation of the rules (i.e. the knowledge acquisition), the maintenance and the update of the rules is in most cases a difficult, time consuming task.

After the events are correlated also with respect to the functional aspect, the resulting events generate automatically a trouble ticket. Fault diagnosis is afterwards performed according to the knowledge of the support staff who still has to correlate trouble tickets reported from users with the automatically generated trouble tickets. However, the task is certainly simplified.

With the introduction of a service management platform, the functional aspect and with this the service dependencies would be handled completely within a service management platform (Fig. 4 (b)). Such correlation would be performed based upon the Service MIB. According to this, an appropriate aggregation of device-oriented events would be possible in order to obtain the status of a service. If a customer reports a problem with his service, and the status of his subscribed service is visualized in a service management platform, fault diagnosis is simplified essentially.

The integration of a service management platform could coordinate also the usage of other tools. Currently, almost every management tool makes its own polling and/or autodiscovery (e.g. SLA tools like Network Health, management platforms, device-specific tools like CiscoWorks) and this in a more or less uncoordinated way. By using a service management platform such in certain extent unnecessary pollings and autodiscoveries could be omitted.

5   Conclusions and Open Issues

Service management is certainly the next upcoming area of research and the next dimension in management. To recognize what are the requirements of these new research topic, aspects of IT service management have been introduced as a basis for a systematic analysis. Some of the requirements have been discussed in more detail and the need for a service management platform has been identified. The paper points to the functionality and the architecture of a service management platform by describing also the integration issue on a correlation scenario.

Beside the mentioned benefits, the usage of a service management platform would also improve proactive management. Due to the knowledge, what impact a component of a service has on the quality of a service, appropriate preventive actions could be taken in advance before serious service degradations would occur.

There are several open issues of IT service management to deal with. To sketch a few, the specification of a service management architecture is one of the most important ones, including tasks of specifying the Service MIB, specifying a framework for dynamic service provisioning etc.

Acknowledgments

The author wishes to thank the members of the Munich Network Management
(MNM) Team for helpful discussion and valuable comments on previous versions
of the paper. The MNM Team, directed by Prof. Dr. Heinz-Gerd Hegering, is
a group of researchers of the Munich Universities and the Leibniz Supercomputing Center of the Bavarian Academy of Sciences. Its webserver is located at
http://wwwmnmteam.informatik.uni-muenchen.de.

References

1. "Service Architecture," TINA baseline, June 1997.
2. "Common Information Modell (CIM) Specification, Version 2.2," Tech. Rep., Distributed Management Task Force, June 1999.
3. P. Bhoj, S. Singhal, and S. Chutani, "SLA Management in Federated Environment," In Sloman et al. [15].
4. L. Lewis, *Service Level Management for Enterprise Networks*, Artech House, 1
   edition, 2000.
5. "Telecom Operations Map," Evaluation Version 1.1 GB910, TeleManagement Forum, Apr. 1999.
6. G. Dreo Rodosek and Th. Kaiser, "Determining the Availability of Distributed
   Applications," In Lazar et al. [14].
7. http://www.caci.com, *COMNET*.
8. IEEE Network, *IP Traffic Engineering*, vol. 14, 2000.
9. G. Dreo Rodosek, Th. Kaiser, and R. Rodosek, "A CSP Approach to IT Service
   Management," In Sloman et al. [15].
10. M. Langer, S. Loidl, and M. Nerb, "Customer Service Management: A More
    Transparent View to Your Subscribed Services," In Sethi [16].
11. H.-G. Hegering, S. Abeck, and B. Neumair, *Integrated Management of Networked
    Systems – Concepts, Architectures a nd their Operational Application*, Morgan
    Kaufmann Publishers, ISBN 1-55860-571-1, 1 edition, 1999.
12. M. Sabin, A. Bakman, E.C. Freuder, and R. D. Russel, "A Constraint-Based
    Approach to Fault Management for Groupware Services," In Sloman et al. [15].
13. G. Dreo Rodosek and Th. Kaiser, "*Intelligent Assistant*: User-Guided Fault Lo-
    calization," In Sethi [16].
14. Aurel Lazar, Roberto Saracco, and Rolf Stadler, Eds., *Proceedings of the 5th
    International Symposium on Integrated Network Management, San Diego*. IFIP,
    Chapman-Hall, May 1997.
15. M. Sloman, S. Mazumdar, and E. Lupu, Eds., *Proceedings of the 6th IFIP/IEEE
    International Symposium on Integrated Network Management, Boston*. IFIP, IEEE
    Publishing, May 1999.
16. Adarshpal S. Sethi, Ed., *Proceedings of the 9th Annual IFIP/IEEE Interna-
    tional Workshop on Distributed Systems: Operations & Management (DSOM'98),
    Newark*, Oct. 1998.

Constructing End-to-End Traffic Flows
for Managing Differentiated Services Networks

Jae-Young Kim[1], James Won-Ki Hong[1], Sook-Hyun Ryu[1], and Tae-Sang Choi [2]

[1] Department of Computer Science and Engineering
Pohang University of Science and Technology
{jay,jwkhong,shryu}@postech.ac.kr
[2] Internet Architecture Team
Internet Technology Department
Electronics and Telecommunications Research Institute
choits@etri.re.kr

Abstract. Differentiated Services (DiffServ), presently being standardized by IETF, is considered to be a promising solution for supporting different service characteristics to different classes of network users on the Internet. The IETF DiffServ working group has defined a general architecture of DiffServ and is elaborating more detailed features. A simple but powerful management mechanism is needed to operate, provision, monitor and control DiffServ networks. Managing end-to-end traffic flows is one of the key components for managing DiffServ networks. Various high-level management functions can be built by using the flow information. In this paper, we present our work on designing a system architecture for managing DiffServ networks using the SNMP framework. DiffServ routers with SNMP agents have been developed, and a management system constructing end-to-end traffic flows has been designed.

1 Introduction

Over the past decade, the number of devices and the number of Internet users have increased at an exponential rate and the network traffics caused by data transfers will continue to rise. While previous network bandwidths were sufficient to carry text-based application data, current network bandwidths are no longer sufficient to handle multimedia, real-time network traffic flows.

Because the increase rate of network bandwidth is much slower than the increase rate of network usage, bottleneck points, where bandwidth is insufficient for network users, are commonly observed. In such situations, every packet competes for access to the bandwidth and the result is packet loss, unexpected delays, and jitter. However, both Transmission Control Protocol (TCP) and Internet Protocol (IP), two network protocols for delivering packets in the Internet, were originally designed in the best-effort service model.

But users' requirements have been changing. Users want to get different service qualities for different types of services they obtain. Integrated Services (IS) [1] with Resource reSerVation Protocol (RSVP) [2] signaling is the first approach to provide such a service on the Internet. RSVP attempts to provide per-flow QoS support assurances with dynamic resource reservation. A flow is defined by the 5-tuple,

A. Ambler, S.B. Calo, and G. Kar (Eds.): DSOM 2000, LNCS 1960, pp. 83 - 94, 2000.
© Springer-Verlag Berlin Heidelberg 2000

consisting of source and destination IP address, transport protocol, and source and destination port. However, since RSVP/IS relies on per-flow states and per-flow processing in every network node, it is difficult to deploy RSVP/IS in large carrier networks like the Internet.

Differentiated Services (DiffServ) is an alternative approach to provide differentiated service qualities to different classes of users. DiffServ uses aggregation of traffics in each routing decision point. Type of Service (ToS) field is used for distinguishing these traffic aggregates. Since the ToS is much simpler than the 5-tuple information, it is easier to implement DiffServ than RSVP/IS [3, 4, 5].

DiffServ applies administrative domain concepts. Within one domain, core routers forward traffics according to the ToS field of traffic aggregates. Between two different domains, there are edge routers which perform classification of flows based on 5-tuple information like RSVP/IS. Since the edge routers mark the ToS field of incoming traffics, core routers do not need to handle complex information.

Although the IETF DiffServ working group has defined several standards for DiffServ, the management aspect of DiffServ is not yet fully standardized. Current standards have defined only the operational aspects of DiffServ. When deploying DiffServ in network nodes, various management functions are needed for remote control of a large number of DiffServ nodes. Possible management considerations are how to configure each DiffServ router, how to change its configuration, and how to monitor or meter traffic each router handles.

Constructing end-to-end traffic flows in DiffServ networks is a key component of managing DiffServ networks. An end-to-end traffic flow consists of a routing path from a source edge router to a destination edge router and performance parameters of packet streams with a given ToS field over the routing path. Various high-level management functions such as bottleneck detection, topology mapping, Service Level Agreement (SLA) monitoring, etc., can be built by using the flow information. Since current management efforts are only focusing on element management of each DiffServ router, the end-to-end traffic flows have to be constructed by using the current element management functions.

In this paper, we propose a method for managing end-to-end traffic flows in DiffServ networks using the SNMP management framework. The IETF DiffServ working group has defined DiffServ MIB for managing DiffServ-enabled network devices. Based on this MIB, we have developed an SNMP agent system that operates in Linux-based DiffServ routers. A central DiffServ manager handles management functions on DiffServ routers with SNMP. The manager constructs end-to-end traffic flows for supporting various high-level management functions. Furthermore, a Web-based DiffServ management console that provides easy-to-use interfaces running in a Web browser is designed.

The rest of this paper is organized as follows. Section 2 explains the architecture of differentiated services proposed by IETF. Section 3 considers the management issues for DiffServ networks. Section 4 describes the detailed constructing methods and applications of end-to-end DiffServ flows and Section 5 shows how to develop a DiffServ management system. Finally, Section 6 summarizes our work and discusses directions for future research.

2   Architecture of DiffServ

DiffServ proposes a basic method to differentiate a set of traffic among network nodes. The method is based on a simple model where traffic entering a network is classified and possibly conditioned at the boundaries of the network, and assigned to different behavior aggregates. Each behavior is identified by a single Differentiated Services Code Point (DSCP).

DSCP is the most-significant 6 bits from the IPv4 Type-Of-Service (ToS) octet or IPv6 traffic class octet. This 6-bit field indicates how each router should treat the packet. This treatment is called a Per-Hop Behavior (PHB). PHB defines how an individual router will treat an individual packet when sending it over the next hop through the network. Being 6 bits long, the DSCP can have one of 64 different binary values.

Four types of PHBs have been defined as standard thus far [4, 6, 7, 8]. They are default, class-selector, Assured Forwarding (AF), and Expedited Forwarding (EF). Table 1 summarizes the standard PHBs and DSCP values accordingly.

Table 1. Standard PHBs

PHB Name	DSCP	Description
Default	000000	best-effort (RFC 1821)
Class-selector	xxx000	7 classes (RFC 2474)
AFxy	xxxyy0	4 classes with 3 drop probabilities (RFC 2597)
EF	101110	no drop (RFC 2598)

A DiffServ-enabled network node has several components for handling DiffServ. Fig. 1 explains five components of DiffServ architecture; classifier, meter, marker, shaper, and dropper [9, 10] in a traffic conditioning block (TCB).

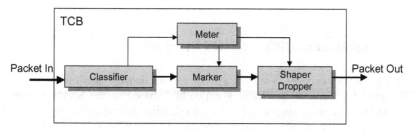

Fig. 1. Basic Traffic Conditioning Block of DiffServ

A classifier selects network packets in a traffic stream based on the content of some portion of the packet header. There are two types of classifiers, the Behavior Aggregate (BA) classifier based on the DiffServ values, and the Multi-Field (MF) classifier based on the value of a combination of 5-tuple information. A meter measures the temporal properties of the stream of packets selected by a classifier. It passes state information to other conditioning actions to trigger a particular action for each packet. A marker sets the DSCP of a packet and a shaper delays some or all of the packets in a traffic stream in order to bring the stream into compliance with a traffic profile. A dropper discards some or all of the packets in a traffic stream in order to bring the stream into compliance with a traffic profile.

DiffServ router is a fundamental DiffServ-enabled network node. The conceptual model and requirements of the DiffServ routers are discussed in IETF [11, 12]. The DiffServ router is considered to have routing component, set of TCBs, queuing component, and configuration and monitoring module that are organized as in Fig. 2.

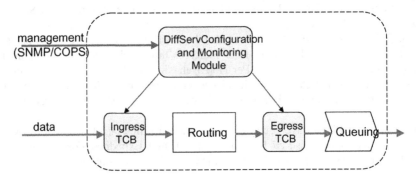

Fig. 2. Conceptual Model of a DiffServ Router

DiffServ-related components are separated from the routing component to simplify the addition of DiffServ capability to the existing router. There is a set of TCBs cascaded both at the ingress point and the egress point. Traffic conditioning can be performed either at the ingress point or at the egress point, or both. Queuing component is a set of underlying packet queues which keep packets before the routers send them out. The management module for DiffServ router can be operated in several ways such as SNMP or COPS [13, 14]. The management module configures TCB parameters and monitors the performance of each TCB. The detailed approach for managing DiffServ networks are explained in the next section.

3   Management of Differentiated Services

Managing DiffServ networks includes a set of various management functions. Current IETF approach for managing DiffServ networks is based on the SNMP framework. The SNMP framework is simple and a de-facto standard for managing Internet-related network devices. We investigate the structure of DiffServ MIB defined by the IETF and provide an overview of on-going efforts to define the MIB.

The IETF DiffServ working group currently suggests an SNMP Management Information Base (MIB) for the DiffServ architecture [15]. The MIB is designed according to the DiffServ implementation conceptual model [12] for managing DiffServ routers in the SNMP framework. The initial draft was proposed on July 1999, with the detailed definitions currently being elaborated and extended in the working group. Table 2 summarizes the primary object tables defined in the DiffServ MIB.

Table 2. DiffServ MIB Structure

Element	Table Name	Description
Classifier	Classifier	list of classifiers
	SixTupleClfr	5-tuple classifier + DSCP value
Meter	Meter	metering parameters
Action	Action	mark / count / absolute drop
Queue	AlgDrop	algorithmic dropper
	Queue	queuing parameters
	Scheduler	shaping parameters

The DiffServ table entries are linked each other with the RowPointer textual convention. RowPointer object is used for pointing an entry in the same or different table [16]. The DiffServ MIB represents a TCB as a series of table entries linked together by RowPointers. With this scheme many different TCBs can be represented in the object tables efficiently. Each table contains several MIB objects to configure, monitor, and modify DiffServ characteristics in a network node. By getting and setting these object values via SNMP, the SNMP manager can control DiffServ-enabled network nodes from a remote location.

However, the current DiffServ MIB is only for managing the characteristics of one DiffServ router. It does not provide a complete network picture of a set of DiffServ routers in one administrative domain. In order to provide such high-level management functions, the current management framework should be extended.

4   Constructing End-to-End DiffServ Flows

We defined a DiffServ flow as a sequence of network packets with the same DSCP value in a DiffServ domain. Every network service provided from a DiffServ network can be represented as a DiffServ flow from a set of source nodes to a set of destination nodes. Possessing information on such DiffServ flows can help understand the current service status. Information on the DiffServ flow consists of two parts: topology and performance. Topology information represents router-to-router connectivity. A path from a set of source edge routers to a set of destination edge routers must be provided. Performance information represents a number of performance parameters of a given DiffServ path. The performance information can be obtained by combining performance parameters of each router in a DiffServ path.

In this section, we suggest a method to create end-to-end DiffServ flows by combining routing information from MIB II and DiffServ performance parameters from DiffServ MIB. The end-to-end DiffServ flow information can be used as a basic component for providing sophisticated high-level management functions.

4.1   Method

A DiffServ flow consists of topology and performance information. Topology information is constructed from routing tables and performance information is constructed from DiffServ MIB values. Constructing end-to-end DiffServ flows thus consists of two phases, as in Fig. 3. First, the topology generator produces the topology information as a linked list of routers and the performance analyzer aggregates performance parameters of each router in the routing path by using the topology information. MIB II and DiffServ MIB are used to construct the information.

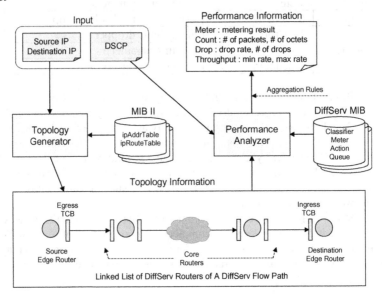

Fig. 3. Construction Process of DiffServ Flow Information

Since each DiffServ router supports routing protocols, the router keeps a routing table containing a list of next hop routers for a given destination IP address. The MIB II has the routing table and a central SNMP manager can retrieve the routing table information to construct a whole routing connectivity map in a DiffServ domain. Two MIB tables, ipAddrTable and ipRouteTable are used to create topology information. The ipAddrTable contains IP addresses of all network interfaces in a router and the ipRouteTable contains the IP routing table that has the next hop host and network interface for a set of destination IP addresses. By combining them we can obtain every source-to-destination routing path. Given a source-destination pair, the topology generator outputs a linked list of DiffServ routers composing a DiffServ flow path.

DiffServ flow performance information is obtained from the DiffServ MIB. Each DiffServ router has performance parameters observed locally. The parameters include metering parameters, counter values, numbers of dropped packets, minimum and maximum rates of packet transmission, and so on. These parameters are calculated and maintained for each DSCP value; that is, the DiffServ MIB of a DiffServ router contains all the performance parameters of DiffServ flows it processes. When a linked list of routers composing a DiffServ flow path is given, the performance analyzer aggregates values of the parameters from each DiffServ router one by one and produces end-to-end performance information of a DiffServ flow.

One important consideration in calculating end-to-end performance information is that the performance parameters contained in the DiffServ MIB in each router do not distinguish packets with different IP source/destination pair. Defined by the DiffServ concept, every core router forwarding packets between the source node to the destination node, only looks up the DSCP value in the header of each packet. Thus performance parameters from DiffServ MIB are for aggregated traffic with a given DSCP value, not for specific traffic flow from a given source to a given destination, which we want to analyze. The traffic flow that we want to distinguish is mixed with other flows with the same DSCP value but with different source/destination pairs.

From this observation, we make rules to follow when aggregating performance parameters. First, absolute values, such as counter values, should be translated to relative values. For example, number of dropped packets should be changed to rate of dropped packets so that the drop rate of a specific end-to-end DiffServ flow can be calculated by accumulating each drop rate in the router list. If there are three routers with 10% drop rates for a specific DSCP flow in the end-to-end routing path, the overall drop rates for the end-to-end DiffServ flow is calculated as 30%. Second, some parameters, such as throughput rates, should be calculated by finding out minimum or maximum values. For example, minimum throughput of an end-to-end DiffServ flow is calculated by finding out the minimum throughput in every router because the end-to-end throughput is bounded by the router with the least throughput.

4.2   Management of DiffServ Flows

By following the proposed method, we can obtain information of a set of end-to-end DiffServ flows in a DiffServ domain. Given a source/destination pair and a DSCP value, topology and performance information of a DiffServ flow from the source to the destination is constructed. Since the flow information gives a network view of DiffServ flows in a DiffServ domain to network administrators, various network management functions can be performed.

● Network topology management

Network topology can be created with the DiffServ flow information. Network connectivity and performance data should be kept in a management system in a certain format. The topology is not static. Numbers of DiffServ flows appear and disappear constantly. Managing the topology should follow such dynamic changes and show the current status.

● Bottleneck detection and rerouting

By analyzing the DiffServ flow information we can find out the location of the traffic bottleneck point. At the bottleneck point, the DiffServ flow cannot satisfy the required throughput. Drop rates go up and the metering result fails. The management system should resolve such occurrences. Rerouting of forwarding paths can be one solution. Routing tables can be modified for high-priority traffic to avoid the bottleneck points.

● Service Level Agreement (SLA) monitoring and reporting

Customers of the DiffServ network always want to know that the quality of service they utilize meets the SLA. Further, service providers want to monitor the service quality they provide to the customers. The service quality measurement turns out to be easy when we have DiffServ flow information. Performance parameters of DiffServ flows from a certain customers' network, which can be monitored and summarized to report the SLA satisfaction.

● Accounting and billing
When the DiffServ is deployed commercially in the Internet backbone, it is necessary for the Internet service providers to keep the usage record of their customers and request fees from them for the amount and quality of the Internet usage. DiffServ flow concepts can be applied to calculate the usage pattern and appropriate amount of fees.

These high-level management issues are under research currently. In the next section, we design a DiffServ management system as an initial framework for supporting the above functions.

5   Developing a DiffServ Management System

In this section, we present a detailed design and on-going implementation processes of a DiffServ management system based on the SNMP framework. The system is currently under development in Linux platforms.

5.1   Design Architecture

The architecture consists of three distinct layers, as depicted in Fig. 4. The three-tier architecture includes a network management system (NMS) client running in a Web browser, an NMS server containing a Web server and DiffServ manager, and network elements performing DiffServ routing and SNMP management.

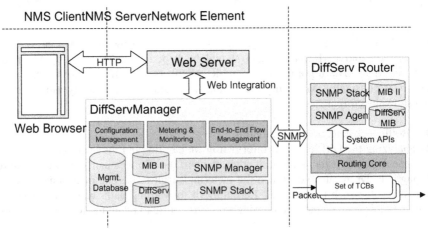

Fig. 4. Design Architecture of the DiffServ Management System

The NMS server is a central server for managing a set of DiffServ routers and providing management interfaces to a set of Web browsers. The Web server located in the NMS server layer has a role to provide a Web-based management interface in Web browsers. The integration of the Web server and the DiffServ manager can be accomplished in various ways such as a basic HTML file access method, a Common Gateway Interface (CGI) method, and a Java applet/servlet method.

The DiffServ manager performs three high-level DiffServ management functions, which are configuration management, metering and monitoring, and end-to-end flow management. The management database is used for storing and retrieving the combined and analyzed data from the MIB II and DiffServ MIB. At the bottom of the DiffServ manager, an SNMP manager communicates with a set of SNMP agents running in different DiffServ routers within a DS domain.

Three high-level DiffServ management functions perform sophisticated and extended management functions. Configuration management function performs remote configuration provisioning. Every DiffServ parameter is determined and enforced via the configuration management function. Metering and monitoring function periodically observes the status of DiffServ routers and compares the results with predefined desirable performance metrics. Such conformance test results are necessary for modifying behaviors of a DiffServ router. Flow management function summarizes all the DiffServ flows in a DS domain and provides the end-to-end DiffServ flow characteristics. The function collects routing tables and DiffServ flow information and constructs overall end-to-end parameters of each DiffServ flow.

DiffServ routers are managed network elements in the design architecture. A DiffServ router contains a routing core module to control a set of TCBs that execute packet forwarding according to various DSCP values, and an SNMP agent module to handle SNMP manager requests for the DiffServ MIB. System-dependent APIs are used to connect the SNMP agent module and the routing core module. The values of DiffServ MIB variables are determined by specific system-dependent system calls. The methods of retrieving and setting DiffServ parameters in the routing core module need not be the same among different implementation architectures.

Within a DiffServ domain, numerous DiffServ routers and DiffServ management clients interwork with each other. The three-tier architecture offers distinct advantages in such environments. One centralized DiffServ manager controls a set of DiffServ routers while providing management interfaces to a set of management clients at the same time. However, by separating the management user interfaces from the manager itself, the DiffServ manager is able to concentrate on management functions and thus the performance of the DiffServ manager can be improved.

5.2   Implementation

Linux, a shareware operating system, supports QoS features in its networking kernel from the kernel version 2.1.90 [17]. The QoS support offers a wide variety of traffic control functions, which can be combined in a modular way. Based on this Linux traffic control framework, W. Almesberger et al. have designed and implemented basic DiffServ classification and manipulation functions required by DiffServ network nodes [18]. The extended DiffServ features are freely available in the form of a kernel patch package [19]. By installing the DiffServ package, a Linux system is able to perform DiffServ router functions.

However, the current Linux DiffServ implementation does not show sufficient management functionality. There is no management architecture and every script setup must be manually configured and modified in local machines. Further, metering and monitoring functions of DiffServ are not fully supported. Our work focuses on this lack of management functionality.

A DiffServ agent is an SNMP agent with MIB II and DiffServ MIB running on the Linux DiffServ router. Basically the agent extracts DiffServ parameters from the Linux traffic control kernel and modifies the appropriate MIB values on the request from a DiffServ manager. The agent also receives management operations from a DiffServ manager and performs the appropriate parameter changes in the Linux traffic control kernel.

The organization of our Linux DiffServ router implementation is explained in Fig. 5. There are two process spaces in the Linux operating system, the user space and the kernel space. Extending from Linux traffic control framework, the Linux DiffServ implementation resides in the kernel space. In the user space, the DiffServ SNMP agent is implemented. Communication between the DiffServ agent and the Linux traffic control kernel is effected via NetLink sockets [20]. The NetLink socket is a socket-type bidirectional communication link located between kernel space and user space. It transfers information between them.

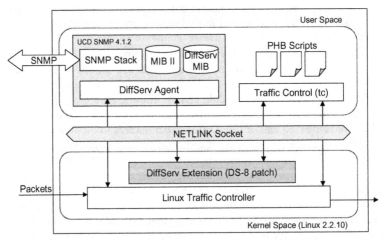

Fig. 5. Organization of Linux DiffServ Router Implementation

The agent has been implemented by using UCD SNMP agent extension package [21]. UCD SNMP 4.1.2 provides the agent development environment. The DiffServ agent uses the traffic control program (tc) or NetLink socket directly for accessing DiffServ parameters in kernel space and manipulates the values of MIB II and DiffServ MIB.

A Web-based DiffServ management system is currently under development in our work. Java programming language is chosen as our development environment because Java applets can be executed in Web browsers very conveniently.

The central DiffServ manager integrated with a Web server is also being developed in a Linux system. It can configure, monitor, and report the characteristics of DiffServ routers and DiffServ networks. A set of DiffServ flow information is constructed by following the method in Section 4 and stored in a PostgreSQL database of version 7.0.2 [22]. For human managers responsible for a DiffServ network, network topology management function and bottleneck detection and rerouting function are in a prototyping stage.

6   Conclusion and Future Work

Differentiated Services (DiffServ) is gaining acceptance as a promising solution for providing QoS support in the Internet. This paper has proposed a method to manage DiffServ using the SNMP framework. Since current research efforts from the IETF DiffServ working group focus mainly on the operational and functional descriptions of DiffServ, a detailed management framework for DiffServ is urgently needed. We have overviewed management concepts for DiffServ by categorizing management operations in the layered architecture and then presented on-going work to define MIB for managing DiffServ-enabled network nodes in the IETF working group.

To overcome current management functional limits and extend the management capability to sophisticated high-level functions, we have suggested a method to construct and maintain end-to-end DiffServ flows by combining MIB II and DiffServ MIB, and showed the applicability of DiffServ flow information. And then we have proposed a DiffServ management system with a flexible three-tier architecture using the SNMP framework. Further, we have developed a DiffServ agent system working in a Linux platform and a Web-based manager system. Management interfaces running in a Web browser enable users to control DiffServ routers conveniently.

In order to improve the proposed DiffServ management system, we are currently working on the following topics.

A systematic method for representing the proposed DiffServ flow information is needed. The proposed construction process must be extended to produce a formal and graphical description of the DiffServ flows. Standardized data formats and graphical representations such as a directed graph with different shapes of vertex are currently being developed.

Scalability of the proposed system should be improved. Current management framework needs constant polling to every router in the management domain. This might not be appropriate, especially in large ISP backbones. To address the scalability problem, the three-tier architecture can be extended to support distributed management functionality with multiple DiffServ managers located in the middle layer. Also instead of polling the routing table, the agent can initiate sending routing change notification to managers by using the SNMP trap method.

Integration with a policy framework is highly recommended. To simplify the system, we have excluded policy management features in this paper, but such a policy framework needs to be integrated with the current SNMP framework for flexible and intelligent configuration and adaptation of DiffServ routers. Future work includes studying the meta-information model for policy representation and designing policy operational modules.

Finally, performance evaluation of the management system we are developing is considered. Because general DiffServ routers handle a huge amount of high-speed traffic, the DiffServ agent must not affect the routing performance of the DiffServ routers. A DiffServ management system needs to be implemented in such a way as to minimize performance degradation.

References

1. R. Braden, D. Clark, and S. Shenker, "Integrated Services in the Internet Architecture: an Overview," IETF RFC 1633, June 1994.
2. R. Braden et al., "ReSerVation Protocol (RSVP) Version 1 Functional Specification," IETF RFC 2205, September 1997.
3. R. Rajan et al., "A Policy Framework for Integrated and Differentiated Services in the Internet," IEEE Network, September/October 1999, pp.36-41.
4. J. Heinanen, "Use of IPv4 TOS Octet to Support Differential Services," IETF Internet-Draft, draft-heinanen-diff-tos-octet-01.txt, November 1997.
5. B. Carpenter and D. Kandlur, "Diversifying Internet Delivery," IEEE Spectrum, Vol. 36, No. 11, November 1999, pp.57-61.
6. K. Nichols et al., "Definition of the Differentiated Services Field (DS Field) in the IPv4 and IPv6 Headers," IETF RFC 2474, December 1998.
7. J. Heinanen et al., "Assured Forwarding PHB Group," IETF RFC 2597, June 1999.
8. V. Jacobson, K. Nichols, and K. Poduri, "An Expedited Forwarding PHB," IETF RFC 2598, June 1999.
9. S. Blake et al., "An Architecture for Differentiated Services," IETF RFC 2475, December 1998.
10. Y. Bernet et al., "A Framework for Differentiated Services," IETF Internet-Draft, draft-ietf-diffserv-framework-02.txt, February 1999.
11. Y. Bernet et al., "Requirements of Diff-serv Boundary Routers," IETF Internet-Draft, draft-bernet-diffedge-01.txt, November 1998.
12. Y. Bernet, A. Smith, S. Blake, and D. Grossman, "A Conceptual Model for Diffserv Routers," IETF Internet-Draft, draft-ietf-diffserv-model-03.txt, May 2000.
13. J. Boyle et al., "The COPS (Common Open Policy Service) Protocol," IETF Internet-Draft, draft-ietf-cops-07.txt, August 1999.
14. R. Yavatkar et al., "COPS Usage for Differentiated Services," IETF Internet-Draft, draft-ietf-rap-cops-pr-00.txt, December 1998.
15. F. Baker, K. H. Chan, and A. Smith, "Management Information Base for Differentiated Services Architecture," IETF Internet-Draft, draft-ietf-diffserv-mib-03.txt, May 2000.
16. W. Stalling, SNMP, SNMPv2, SNMPv3, and RMON 1, 2, 3rd Edition, Addison-Wesley, 1999.
17. S. Radhakrishnan, "Linux – Advanced Networking Overview – Version 1," a technical paper of Department of Electrical Engineering and Computer Science, University of Kansas, August 22, 1999.
18. W. Almesberger, J. H. Salim, and A. Kuznetsov, "Differentiated Services on Linux," IETF Internet-Draft, draft-almesberger-wajhak-diffserv-linux-01.txt, June 1999.
19. W. Almesberger, Differentiated Services on Linux, Internet Web site, http://lrcwww.epfl.ch/linux-diffserv/.
20. ITU-T Recommendation M.3010, "Principles for a Telecommunications Management Network," 1996.
21. UCD-SNMP homepage, http://ucd-snmp.ucdavis.edu/.
22. PostgreSQL homepage, http://www.postgresql.org/.

Platform Architecture for Internet Service Delivery and Management

Gaëtan Vanet, Motohiro Suzuki, Tôru Egashira, and Yoshiaki Kiriha

Computer & Communication Media Research
NEC Corporation
1-1, Miyazaki 4-Chome, Miyamae-Ku,
Kawasaki 216-8555, Japan
{vanet,motohiro,egashira,kiriha}@ccm.cl.nec.co.jp

Abstract. Through the current explosion of the Internet, service consumers expect advanced Internet services become more customizable and easy to use. So, Service Providers must be able to deploy and manage such services, reducing the time to market of its services while drastically improving the Service Quality. However, the current Internet technology is lacking when it comes to developing and deploying new services, integrating together service offerings. This paper proposes a platform architecture for Internet Service delivering and management. We also introduce the concept of "basic service component" as an extension of the notion of building block defined by the TeleManagement Forum and as a way to build business services in a flexible manner. Our proposal makes full use of the Telecommunications Information Networking Architecture (TINA) Business Model with some adaptations to fit with our component-based approach.
Keywords. Basic Service Component, Building Block, Composite Internet Service, TINA Business Model.

1 Introduction

Through the current explosion of the IP technology, service customers expect the Internet to become a multiservice worldwide platform, supporting a full range of advanced IP services including video conferencing, application rental or network based training service. Besides, data services are expected to grow to around 70 percent of the network bandwidth use within the next 5 years. However, this multi-billion market has significant implications for Service Providers.

Service Providers have to comply with different requirements of the market: as customers have now the choice between different providers, the competitive context of the Telecommunication environment requires the Service Provider to reduce the time to market of its services. Moreover, service customers now expect advanced IP services become more intelligent and more customer-specific, e.g. adaptable to each individual requirements while drastically improving the Service Quality. This service customization entails different service compositions and service presentations, proposing, for instance, different types and contents of user interface to customers. Furthermore, the behavior of customers has drastically changed and the service delivery must fit with the variety of customer service access capabilities which can include laptops, PDAs and mobile phones. Finally, the service fulfillment process must be done dynamically, according to the requirements of the users and the conditions of the network. The finality of this evolution is to make advanced IP services "consumable", e.g. pay-per-use, and easy to use like the telephone today.

A. Ambler, S.B.Calo, and G. Kar (Eds.): DSOM 2000, LNCS 1960, pp. 95 - 106, 2000.
© Springer-Verlag Berlin Heidelberg 2000

The main challenge for service providers is the management of such new services which may rely on mechanisms like policy-based management, Service Level Agreement (SLA) management, Customer Network Management (CNM), and automatic remote network configuration and monitoring. In this context, the current Internet technologies are lacking when it comes to developing and deploying new services, integrating together service offerings. Even if this is a complicated challenge, the combination of technologies like Active Networking, Workflow Engine and Application Components can provide a suitable solution to this challenge.

The following section introduces the evolution of the Telecommunication Business model and discusses why the active network, the application components and the workflow engine technologies can constitute a suitable combination to deploy and manage advanced IP services. Section 3 introduces the concept of basic service components and IP service modularity. Then, we presents our proposal of platform architecture in section 4. The service fulfillment and assurance processes are detailed in section 5. Our implementation relying over the NEC front-end middleware is introduced in section 6. Section 7 concludes this paper and presents topics for further research.

2 Evolution of the Telecommunication Business Model

As we aforementioned, business models of the past are not suitable anymore and must be updated to fit with the evolution of the Telecommunication Business Environment. Indeed, new business models try to end up with the monolithic system wherein the network provider is connectivity provider, content provider and retailer at the same time. Figure 1 represents the Business Model defined by the Telecommunications Information Networking Architecture Consortium (TINA-C). For further details about the role of each business role and business relationship, the reader should refer to [10].

In the current and popular model of the Internet environment, four actors have unique roles in supplying or consuming advanced IP services: the service consumer, the Application Service Provider (ASP), the Network Service Provider (NSP) and the Service Portal. The ASP plays the role of the 3rd Party Service Provider defined in the TINA-C model. It focuses its activity on providing a growing range of advanced IP services including application rental or video conferencing. The ASP can be either a service logic provider or a content provider or both. It develops the code necessary for the implementation of these IP services that will be deployed within the network. The NSP, the connectivity provider, sells added-value services analogous to the multitude of enhanced voice services that are offered to customers along with basic voice transport. We believe that the NSP must provide customers with services going beyond the classical network services in order to survive in this new telecom environment. Finally, the Service Portal plays both the role of service retailer and broker, providing a portal to individual and corporate customers. Yahoo or Altavista are good examples of popular Service Portals. According to this Business Context, the dynamic insertion of code, for both the service logic and the service management, is necessary to provide customers with enough service flexibility. Furthermore, we must increase the use of functional service components to enable advanced IP services to be more customizable and code more easily re-usable.

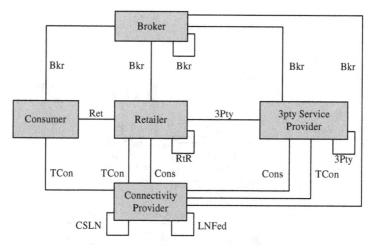

Figure 1. TINA Business Model Relationships Types.

The Active Network technology [6,13] enables the insertion of new services in the network. As Juhola and al. mentioned [4], Active Networking can improve the ability of network providers to penetrate the lucrative Internet-based markets by bringing the following benefits: a) inherently mobility of users, b) more available services based on library of service components, c) additional flexibility to facilitate fast service introduction and enhancements for complex services. We also anticipate that 3rd party development of value added services will arise, providing customers with an increased range of application level services in response to their various needs. In addition, the work of the TeleManagement Forum (TM Forum) to provide guidelines to design application architectures using the concept of Building Blocks [11] provides services with better modularity. This application components-based approach can reduce the time-to-market while also proposing re-usable code and customizable services. In other respects, the Workflow technology can link these service blocks together in a dynamic and flexible manner. The aim of this paper is to propose a platform architecture to deploy and manage Internet services. In this architecture, Internet services are composed of service components and the workflow engine is presented as one technology to link these components together. Our proposal for a new platform architecture for delivering and managing component-based IP services is compliant with the efforts of TINA-C and TM Forum to define truly open architectures.

3 The Service Modularity

As we mentioned in the introduction, IP services must become more customizable, rather viewed as an integration of basic service components than a static and unified service, common to all customers as it is the case today [1,5,14]. In fact, the concept of *component* should be rather considered as an evolution of the object-oriented development methodologies to enable better system integration than its direct competitor. The CORBA Components [9] and Enterprise JavaBeans [3] technologies are the main actors driving this evolution towards component based systems. In these emerging standards, a component is a reusable program block that provides a business logic and can be combined with others in a distributed manner to build an application.

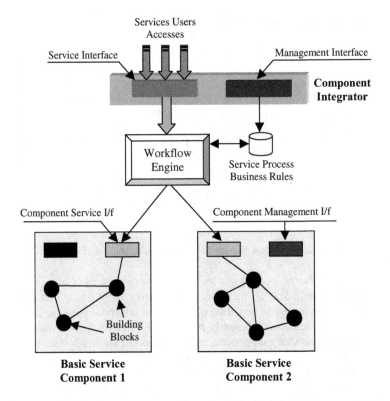

Figure 2. Model of Component-Based IP Service.

In our proposal, we indeed make the distinction between the views of the service that the customer and the provider may respectively have. From the customer's point of view, the service is the business product he pays for and can be confined to the role of Service Access Point (SAP). For the service provider, the service also includes all the service components which must be implemented to provide customers with this business product. In this document, we called them "basic service component". Each service component is re-usable and sharable between different business services.

Figure 2 represents the model of a component-based IP service. Thus, a business service is a composition of basic service components. The integration of these service components is done by the *Workflow Engine*. The Workflow Engine schedules the flow of messages between the different basic service components. Some rules are associated with this workflow system to detail the flow of data between the different basic components providing the business service. Currently, different research projects [7,14] are investigating the use of workflow system to enable management application integration. However, not so many efforts are done towards the integration of service components. In our proposal, the service users access the service through the *Component Integrator* (CI). The Component Integrator is composed of a service interface and a management interface. The *service interface* is intended for service customers to use the service. It provides customers with the list of all the methods related to the specific service. On the other hand, the *management interface* enables the manager to set-up, supervise and control the service. For instance, he can update the rules defining the workflow process. The Component Integrator also contains service meta-data defining the type and the location of the basic service components which must be assembled to build up the corresponding business service.

A Basic Service Component is also composed of two different interfaces: a management and a service interface. The component *management interface* enables the component manager to start/stop the component, get the number of customers connected, the transaction rate or the computing resource usage. On the other hand, the component *service interface* lists the capabilities related to the specific service logic provided by the service component. In the case of a basic service component proposing a billing service, the service interface must include capabilities to configure the component, manage a client billing profile, and so on. To make an analogy with the *building block* concept elaborated by the TM Forum in the Telecom Integration Map document [11], the basic service components are both composed of different building blocks with functionalities that may belong to different computing tiers. They can be implemented as a *process component* based on the CORBA Components specification [9].

4 Platform Architecture for Internet Services

As we explained in the previous section, this is very promising to design a service platform architecture that considers a service as a composition of basic and reusable service components. However, the deployment and the management of such services is a difficult task for Service Providers. This section details our proposal of platform architecture for delivering and managing these composite Internet services.

4.1 Overview

The authors of the FlowThru project [7] has proposed a mapping of the TM Forum Business Processes onto TINA Business Roles. In their mapping, the operational processes defined by the TM Forum are assigned to each business role of the TINA Business Model. Their model was defined to be applied to the Telecommunication Environment but can be modified to fit with our component-based Internet service approach.

So, we propose in this paper a new mapping to fit with the requirements of the IP services environment and with our composed-based services approach. This mapping is depicted in figure 3. We kept the same business roles as defined by the TINA Consortium. However, we modified the business processes applied to each business role. First, we consider two types of 3rd party service provider: the first one is a Basic Service Component Provider which provides the service logic while the second one is a Content Provider, focusing its activity only on the content of the service. However, the TINA Business Model defines the business role of Connectivity Provider that we do not have in figure 3. In fact, we are considering one more hierarchical level and the connectivity issue is masked. On the other hand, we introduced new business processes into the 3rd party service provider role to manage the basic service components (BSC) or the content elements. The broker, consumer and retailer business roles are as defined by the TINA Consortium. Actually, the IP services world requires the consideration of the computational aspect of the service components, aspect which is cur-

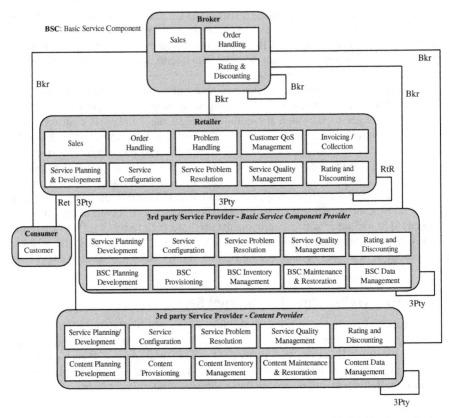

Figure 3. Mapping of the TM Forum Business Processes onto TINA Business Roles Applied to the Component-Based IP Services Scenario.

rently missing in the TINA Business Model. This feature has been taken into consideration for the elaboration of the model depicted in figure 3.

4.2 Business Roles Description

The previous section introduced the overview of our service platform proposal. As our proposal focuses on the deployment and management of component-based services, we are going to explain into further details the functions of the Retailer and 3rd Party Service Provider – Basic Service Component Provider presented in figure 3. In this section, we define components composed of one or several Business Processes.

4.2.1 The Retailer

As a Service Contractor, the retailer guarantees the service by providing the necessary Service Access Points and fulfilling the conditions defined in the SLA, signed with the service consumer. It also collects the accounting information for the purpose of billing. It manages the service user profiles, processing policies and achieves the authorization prior to service usage. Indeed, the service retailer is responsible for the business processes specified in figure 3. These have been defined by the TM Forum [12] but we explain in this section their function considering our composite Internet services approach.

The *Sales* process component comprises the functions necessary for the interactions with the customer, the sale of the services and the translation of customer requests into right actions. The classical HTTP might be the suitable support for the interactions with clients. It also achieves the management of the Service Access Points spanned throughout the network and maintains the repository of all the services which can be provided to clients.

The *Order Handling* and *Service Configuration* process together comprise all the management functions needed in order to define service offerings, administer customers and users, and manage the details of service provisioning. In our proposal, these processes span the service and basic service component layer. According to the service chosen by the customer, it defines the list of basic service components which must be implemented to provide this business service and contacts the right 3rd party service providers to ask them to deploy these components.

The *Customer QoS Management* process checks if the SLA signed with the service consumers are respected. In our case, this process supervises the end-to-end service, at the level of the Component Integrator. It does not check the QoS at the basic service component level. It manages all the data related to clients including SLA requirements, the list of services he can access and the corresponding QoS. It monitors, manages and reports the quality of service. The *Service Quality Management* process monitors the service at the service class level. Indeed, this process supervises the Quality of Service and if the service levels are not met, it must achieve some reconfiguration. In our composite service approach, the retailer may decide to contact 3rd party service providers to deploy new service components.

The *Problem Handling* process achieves the interactions with the service client concerning the service problems. If the customer notifies a SLA violation, this process contacts the *Service Problem Resolution* process. This latter first defines the basic service components composing this service. Then, it contacts the different basic service providers to notify them about this problem. These must then find the solution to solve this service trouble.

The *Service Planning & Development,* the *Invoicing/Collection* and the *Rating and Discounting* process are similar as those defined in the TM Forum Telecom Operations Map document [12]. The *Invoicing/Collection* process manages the billing while the *Rating and Discounting* process applies the basic rating rules and the discounting policies if the Service Level Agreements were not met.

4.2.2 Basic Service Component Provider

The Basic Service Component Provider business role is a 3rd party service provider. Its main function is the provisioning and management of the basic service components rent to service retailers. Indeed, the Basic Service Component Provider is responsible for the business processes specified in figure 3. These have been defined by the TM Forum [12] but we explain in this section their function considering our composite Internet services approach.

The *Basic Service Component Provisioning* process manages the provisioning of the basic service components required by service retailers. This deployment of basic service components in the network is done according to the requests of retailers or 3rd party service providers. The *Basic Service Component Inventory Management* is involved in the code management of the service components. The basic service components repository can be distributed throughout the network and the code of these components can be stored in code servers. The *Basic Service Component Planning/Development* process deals with the development and the acceptance of new service components. It can also contact other Basic Service Component Providers to provide better offers to service retailers.

The *Basic Service Component Maintenance & Restoration* process monitors the service components installed and running. If it detects abnormalities, it can re-act based on rules. For instance, it replaces service components or deploys new ones. The *Basic Service Component Data Management* process gets a collection of usage data and events necessary for the purpose of service components performance and service analysis. This process may gather information from agents monitoring the "health" of basic service components or directly through events emitted by the basic service components themselves.

The 3rd party service provider can have business relationship with other Basic Service Component Provider to provide service retailers with a more complete offer. In this case, the *Service Configuration*, the *Service Problem Resolution*, the *Service Quality Management,* the *Service Planning/Development* and the *Rating and Discounting* process have the same function as specified above for the service retailer but related to these "outsourced" service components.

5   Internet Service Management

The Service Management includes the installation, running, supervision and removal of a service in a networking infrastructure. The TM Forum Telecom Operations Map [12] has defined three major steps in the "Flow-Through" process followed by a service provider to provide clients with services: fulfillment, assurance and billing. The fulfillment and the assurance steps are detailed in this section.

5.1   Service Fulfillment

The *Service Fulfillment* process is responsible for timely delivering what the customer ordered. The design, deployment and activation of advanced IP services require multiple steps. Firstly, the 3rd party service providers contact the service retailers to indicate the basic service components they can provide along with the corresponding SLA. Then, the service consumer connects the network via classical dial-up access and connects the service retailer. This latter defines the Service Home Page (SHP) specific to this customer. This SHP can be considered as a user interface, Web page or Java applet, listing all the services the consumer may access. Afterward, the client chooses the service he wants to access along with the required SLA. The Order Handling process of the service retailer checks the rights and the credit of this customer and creates his session account. After that, the Service Configuration process defines the list of the basic service components which must be implemented to provide the service. Next, the service retailer contacts the 3rd party service providers which sell these components and specifies the location where these components must be installed.

The Basic Service Component Provisioning process of the 3rd party service provider locates the basic service components that are already implemented in the network. If the right components are not already in place, it deploys them. Actually, each basic service component has its own SLA signed between the service retailer and the

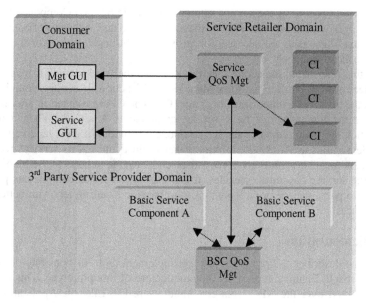

Figure 4. Example of Service Assurance Process.

3rd party service provider. Afterward, the service retailer deploys and activates the service interfaces the service consumer will access to use his required business service. The service profile business rules of the workflow engine must also be updated to take into account this service provisioning phase.

5.2 Service Assurance

In the current state of the Telecommunication environment, Service Providers wants to provide their customers with SLAs, stating the obligations defined between the provider and the consumer. The aim of the *Service Assurance* process is to control that the conditions specified in SLAs are respected and if it is not the case, it must find the source of this "trouble" and correct it. Besides, the Service Assurance will have to apply the service discount policy to the customer according to the level of the "trouble" damage. In order to present the communication between the different entities of our platform architecture for IP services management, we consider the scenario depicted in figure 4. In this example, a business service is composed of two basic service components, called A and B, provided by a Basic Service Component Provider.

Two GUIs have been defined in the Consumer Domain. The purpose of the Management GUI is to enable the service client to check the Quality of his Service. The Service GUI is the interface provided by the Service the client is accessing. The service retailer plays the role of Quality and Accounting Manager concerning the services provided to customers while the 3rd party service provider plays the role of Quality Manager and Accounting Manager concerning the different basic service components he provides to the service retailer. Periodically, the Basic Service Component QoS Manager checks if the performance of its basic service components A and B respects the SLA signed with the service retailer. Indeed, it handles the deployment, the execution and the management of its service components. When it detects a "trouble", a "trouble ticket" is sent to the service retailer. The Basic Service Component Provider also exchanges with the service retailer the accounting data for the use of basic service components. These data are necessary to provide the billing function for the usage of the service the consumer does.

On the other hand, the service retailer achieves the "end-to-end" management of the service through the Service QoS Management function. It deploys, monitors and manages the different Component Integrator (CI). The CI represented in gray in figure 4 achieves the integration of basic service components A and B. When it receives "trouble tickets" from the basic service component provider, it must be able to handle the problem at the service level. It implies that it must find out its services which are degraded and provides solutions. It may also send "trouble tickets" to the Management GUI of the service consumer. Besides, it can use basic service components provided by other 3rd party service providers in order to provide its customers with better Quality of Service.

6 Implementation

This section describes the prototype we are implementing to demonstrate the validity of our approach. Figure 5 represents the architecture of this prototype. This prototype is implemented over the NEC front-end middleware [2]. Like the Application Layer Active Networking proposal [8], the front-end middleware enables a good integration with existing IP networks by simply overlaying the basic network infrastructure. In fact, the front-end middleware enables the dynamic deployment of piece of code, called front-end, in active nodes located at the edge of the network. A front-end is devoted to the preliminary processing of data going from clients to servers. The use of the front-end concept brings different benefits: putting the front-ends at the edge nodes of the network reduces the service response time; front-ends provide customers with more user-friendly and customized interfaces; besides, application servers can subcontract specific tasks to these front-end elements and focus their resources on fundamental tasks; finally, front-ends limit the network traffic by only forwarding to servers the data they cannot process. It enables the network to provide some part of a business service, going beyond the "traditional" approaches like data caching or load balancing which only provides data.

This prototype can be considered as one possible implementation of the generic platform architecture detailed in the previous sections of this document. Actually, our prototype is composed of the following entities:

- The *Service Client* has the consumer business role. We assume these users access the network through Points of Presence, or POPs, and use the classical Web browser as service interface. Actually, the ubiquity of Web browsers makes them the de-facto standard for service interface;

- The Service Broker is playing the role of service retailer. Its function is to sell and manage business services to both individual and corporate customers. At a computational level, the function of Service Broker can be replicated over several nodes of the network;

- Finally, the Service Component Provider and the Content Provider play the business role of 3rd party service provider. The Service Component Provider provides the basic service components necessary to build up the service logic while the

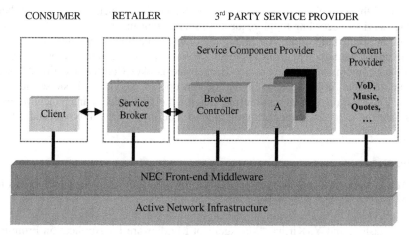

Figure 5. Overview of the Implementation Architecture.

Content Provider provides the content of the service. The business of the Service Component Provider is to develop and sell basic service components to service retailers. The *Broker Controller* deploys and manages the basic service components. Each 3rd party service provider has its own Broker Controller.

The broker business role defined in the TINA business model has not been implemented in our platform prototype. Actually, the number of actors of the telecom environment is currently limited and the retailer often also plays the role of broker at the same time. In a future work, we plan to separate these two business roles by implementing two separate entities. However, the front-end middleware has defined a White Page service that we use to implement our proposal. Besides, the use of the front-end middleware in our implementation enables the dynamic deployment of basic service components code within the network. In this context, the different entities of our proposal and the basic service components are implemented as front-end, enabling an on-demand and easy deployment of the service in the network. Besides, the authors are also investigating the Enterprise JavaBeans and CORBA Components technologies to improve our implementation.

7 Conclusions

This paper has described a platform architecture for deploying and managing component-based Internet services. In this context, a business service is viewed as an integration of basic service components. These components are reusable and shareable between business services. Our aim was to provide an enhancement to the concept of building blocks in order to provide service providers with platform for advanced Internet services management. The main philosophy of our generic platform architecture proposal is to enable service retailers to deploy a new service for immediate use by any consumer in the TINA system without standardizing the services with other retailers. Our platform architecture proposal makes the use of three strong and promising technologies: programmable networks, application components and workflow engine.

The aim of this combination is to allow the dynamic deployment of component-based Internet services according to the requests of service consumers.

In the immediate future, we intend to complete the implementation of our prototype to check the validity of our approach. We also plan to define in more details the functional components of the Service Broker and Broker Controller entities. Finally, we want to make further research about the concepts of "service", "basic service component", and the degree of possible mapping of the IP service management architectures onto the Telecom standards and guidelines defined by the TINA-C and TM Forum.

References

1. Active Network Composable Services Working Group, "Composable Services for Active Networks", September 1998,
    http://www.cc.gatech.edu/projects/canes/compserv/cs-draft0-3.pdf
2. T. Egashira and Y. Kiriha, "Management Middleware for Application Front-ends on Active Networks", in Proceedings of NOMS 2000, Hawaii, April 2000.
3. Enterprise JavaBeans, http://java.sun.com/products/ejb/index.html
4. A. Juhola, I. Marshall, S. Covaci, T. Velte, M. Donohoe and S. Parkkila, "The Impact of Active Networks on Established Network Operators", in Proceedings of IWAN 1999, Berlin, July 1999.
5. G. Kar, A.Keller, S. Calo, "Managing Application Services over Service Provider Networks: Architecture and Dependency Analysis", in Proceedings of NOMS 2000, Hawaii, April 2000.
6. U. Legedza, D.J. Wetherall and J. Guttag, "Improving the Performance of Distributed Applications Using Active Networks", IEEE Infocom, San Francisco, April 1998.
7. D. Lewis and al., "Implementing Integrated Management Business Processes uing Reusable Components", January 2000,
    http://www.cs.ucl.ac.uk/research/flowthru/content/finalrpt/flowthru-rpt.pdf
8. I. Marshall, M. Fry, L. Velasco and A. Ghosh, "Active Information Networks and XML", in Proceedings of IWAN 1999, Berlin, July 1999.
9. OMG TC Document orbos/99-02-05, "CORBA Components: Joint Revised Submission", March 1, 1999.
10. Telecommunications Information Networking Architecture Consortium (TINA-C), "TINA Business Model and Reference Points – Version 4.0", May 22, 1997.
11. TeleManagement Forum, "Generic Requirements for Telecommunications Management Building Blocks - GB 909 (part 1)", Evaluation Version 2.0, September 1999.
12. TeleManagement Forum, "Telecom Operations Map - GB 910", Evaluation Version Release 1.0, October 1998.
13. D.L. Tennenhouse, at al., "A Survey of Active Network Research", IEEE Communications Magazine, Vol. 35, No. 1, January 1997, pp.80-85.
14. P. Wade and T. Richardson, "Workflow – A Unified Technology for Operational Support Systems", in Proceedings of NOMS 2000, Hawaii, April 2000.

Middleware Platform Management
Based on Portable Interceptors

Olaf Kath[1], Aart van Halteren[2],
Frank Stoinski[1], Maarten Wegdam[3], and Mike Fisher[4]

[1] Humboldt-Universität zu Berlin
{kath,stoinski}@informatik.hu-berlin.de

[2] KPN Research
A.T.vanHalteren@kpn.com

[3] Lucent Technologies - Bell Labs Twente
wegdam@lucent.com

[4] BT Advanced Communications Technology Centre
mike.fisher@bt.com

Abstract. Object middleware is an enabling technology for distributed applications that are required to operate in heterogeneous computing and communication environments. Although hiding distribution aspects to application designers proves beneficial, in an operational environment system managers may need detailed information on information flows and the locality of objects in order to track problems or tune the system. Therefore, hooks are required inside the processing core of the middleware to obtain inside-information and to influence the processing of information flows. We present the use of portable interceptors for the management of CORBA as well as COM/DCOM middleware. Management information is structured in a middleware technology independent way, using XML for representation. Our approach shows two aspects of "management transparency": application designers are not burdened with designing management functionality, and system managers can manage CORBA and (D)COM from a single set of management tools.

1 Introduction

This paper gives an overview of some intermediate results from the EURESCOM Project 910, Technology Assessment of Middleware for Telecommunications. The project builds on the claim that Public Network Operators (PNOs) and Service Providers can benefit from distributed object technologies and middleware platforms. The purpose of the collaboration is the assessment of middleware technology as a means of providing large scale software infrastructures, suitable for wide range of telecommunication services. The approach is to assess middleware technologies by means of hands-on experience gained through actual experiments.

A key element for a large-scale software infrastructure is the operational management of the middleware and the software components that constitute the infrastructure. Management of middleware is essential for service providers to operate large-scale software infrastructures. These infrastructures will unavoidably consist of multi-vendor software solutions. Operational management of heterogeneous multi-vendor software infrastructures requires the development of middleware management concepts, such as relevant management information and management policies.

A. Ambler, S.B. Calo, and G Kar (Eds.): DSOM 2000, LNCS 1960, pp. 107- 118, 2000.

These concepts must be represented in a concrete, but middleware platform technology independent notation. Section II explains our approach towards developing middleware management concepts and their representation. In addition, middleware management information must be obtained in a standardized, portable way to hook management tools into the core of middleware software infrastructures, e.g. hooks into ORBs and object services. Examples of such hooks are CORBAs Portable Interceptors and COM/ DCOM interceptors. Due to the unavailability of a standard and products supporting portable interceptors during the project lifetime, our own specification was developed for CORBA and (D)COM platforms, prototypically implemented in a multi vendor CORBA environment and contributed to the OMG standardisation process[7]. Section III explains our interceptor design.

All presented middleware management concepts were implemented as prototypes in a multi-vendor CORBA environment; some of the implementation concepts are presented in Section IV. Conclusions and future work are presented in Section V.

2   Middleware Management Concepts

2.1   Management Transparency and Management Information

Object middleware offers the so-called distribution transparencies [5]. This means that the application developer does not have to be aware of issues like the location of objects, the user programming languages used, the used network, available transport protocols etc. He can simply focus on the business logic of the software component he is developing, reducing time-to-market and decreasing development costs. Along the same lines, we believe that the middleware should also offer management functionality, and that this management functionality should also be transparent for the application developer. This management transparency does not only have to be offered to the application developer, but also to the system manager. A system manager has to manage an application consisting of software components running distributed objects which have been developed independently using different technologies. He should be able to manage using a single management platform. Management information in a distributed system can roughly be divided into three different categories; application specific management information, information which has to do specifically with a middleware platform, and operating system and network resources specific management information. Only the middleware platform management information is within the scope of this paper. Figure 1 depicts the various information categories, and further divides the middleware management information into object related information, request related information, message level information and network level information.

2.2   Multi-domain Management

Our focus in this work is on middleware for large-scale distributed systems crossing organisational boundaries. It is widely recognised that business relationships are rapidly becoming much more dynamic. Increasing automation in cross-business processes means that enterprise systems need to work together in heterogeneous groupings, the details of which are unknown to the designers of any of the systems involved. A similar situation is typical in any large organisation, where there is a need for applications designed and built independently to cooperate effectively in unforeseen situations.

Management in these changing environments presents a number of new problems. There are multiple management domains with multiple uncoordinated points of control, each making independent design decisions and using a range of middleware technologies. Centralised decision making is not a viable option for large systems. Global coor-

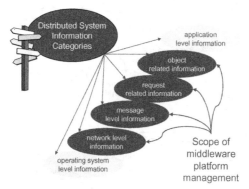

Fig. 1.   Scope of Middleware Platform Management Related Information

dination is not possible and the response of the system will be the result of a collection of autonomous actions. However, end-to-end management is required so there is a need to exchange relevant management information and policies between the interacting systems.

Information exchange between domains requires common information models or, at least, common information modelling principles and syntax. Here we make the reasonable assumption that an approach based on Events (for monitoring the state of system components) and Policies (for expressing the desired behaviour of system components) will be applicable to the management of a range of different middleware technologies. Our principal focus here is on mechanisms to support management of heterogeneous middleware platforms. Management information and policies should therefore be represented in a platform technology independent way. This is essential for information that is relevant across multiple platform technologies. For specific platforms specialisation of management information is anticipated.

XML [4] is becoming the de facto standard for information representation and exchange, particularly where the information must be automatically processed. It allows users to define representations specific to their own applications with a well-defined formal syntax. XML meets our requirement for a syntax which is not tied to a specific middleware technology but which has the necessary flexibility to represent a very wide range of information. Our approach, therefore, is to define management concepts using XML DTDs (Document Type Definitions) and then to apply a mapping to platform specific formats.

Figure 2 illustrates this approach. A prerequisite is a common understanding of management concepts and processes which have relevance wider than a single product (across different CORBA implementations, for example), or middleware technology (across CORBA and (D)COM, for example). These concepts are then expressed in XML DTDs or Schema to provide reference representations of information which has a consistent meaning in multiple systems. Ideally these representations would be standardised but where this is not possible, automated transformation can be used.

The next step is to map in a well-defined way to specific technologies, such as for example CORBA and (D)COM. This approach does not impose restrictions on the message coding, protocols or management mechanisms used within a specific technology or product. This is similar to the way the CORBA specifications allow interoperability but also diversity in implementation. There is the capability of expressing information in a form that can be unambiguously understood in another

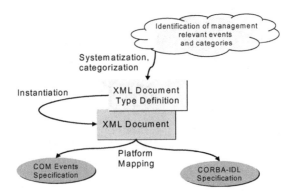

Fig. 2. Platform Technology Mapping of Management Relevant Events

technology domain when required. Although conceptually this is accomplished via the reference XML representation, there is no need actually to generate an intermediate XML document. Similarly, although it is a useful lowest common denominator, particularly in wide area networks, the XML-based approach does not imply the use of http as a transport protocol. The representation syntax and the protocols for information exchange are independent of each other.

3   Plug-In of Management Services into the Middleware Core

Management of a multi-vendor middleware infrastructure requires a standard way for obtaining management information and controlling the infrastructure (i.e. application of policies). This requirement in turn implies the need for a technology which allows management components to interact with the core components of a middleware platform in a portable, product independent manner. By the core of a middleware platform we mean runtime components that provide all the necessary basic capabilities for the interaction between distributed objects. Examples of such core components are object request brokers (ORBs) running on nodes belonging to a CORBA based platform, or containers that host Enterprise Java Beans (EJB Containers).

Current approaches for providing such standardized interaction hooks are CORBA Portable Interceptors [9] and Interceptors in COM+. During the project, both approaches were evaluated for their applicability for platform management purposes.

3.1   CORBA Interceptors

Interceptors were introduced into CORBA to provide a way for security mechanisms to interact with the ORB without the ORB having to be explicitly aware of the mechanisms. Interceptors clearly have potential application in a number of other areas, in particular management of CORBA systems, and several ORB vendors have provided interception mechanisms to increase the flexibility of their products. It turns out, that Portable Interceptors can be used to collect dynamic management information of a CORBA runtime system and to verify if the system complies with the management policies. However, interceptors in CORBA 2.2 are underspecified and not portable. The OMG issued a Portable Interceptors RFP [9] for 'a portable definition of interceptors so that system services and users may "plug into" ORB processing at particular points'. Due to the unavailability of an adopted portable interceptors specification, we have developed our own architectural model and a detailed specification, and contributed that to the standardization process within the OMG [7].

Based on an abstract ORB processing model, intercepted actions or interactions can be categorised into conceptual different abstraction levels, mainly object level interceptors, request related interceptors (request level and message level interceptors) and network level interceptors. For reasons of brevity, we present here only the conceptual model and a number of design decisions we have taken, and not the whole specification.

Object Lifetime Related Interception

We have identified the need to monitor and control the lifecycle of objects, proxies and object adapters. These lifecycle states can be intercepted just before and/or just after the lifecycle state changed, using an Object Adapter (OA) interceptor.

We focus here on the Portable Object Adapter (POA), since we do not consider it worth the effort to also include the deprecated Basic Object Adapter. Information that can be monitored or changed with an OA-interceptor includes the IOR template, the name, the parent POA and the policies that will be used. In a POA context there are four state changes for an object: creation, activation, deactivation and destruction. Potentially each of these four can be intercepted just before and just after the state change.

At the client side we identified a need to intercept the creation of a proxy. The term proxy refers to the client side ORB internal representation of an object referenced by an interoperable object reference (IOR). With a request interceptor it is possible to discover a connection between a client and a server as soon as it is actually used (at the first request). A proxy interceptor however allows the discovery of a logical connection between a client and a server before it is actually used, or even if it is never used. Available information is the IOR that was used to create the proxy, and the resulting object (i.e. proxy).

Request Related Interception

Request related interceptors are invoked during the ORBs processing of operation invocations. During information modelling, we identified several interception points and the information available at each point.

There are four interception points which directly correspond to activities of an ORB or POA during the processing of a particular operation invocation:

- The client side ORB receives the operation invocation from a client co-located to it. For this activity, we identified the corresponding interception point *ClientPreInvoke*. The client side ORB communicates the operation invocation in some form, corresponding to a particular interoperability protocol, to the target object, using either a network connection or some local invocation paradigm, if the target is in the same capsule.
- The target ORB receives the operation invocation and identifies a servant which implements that operation. Before the servant is invoked, we identified the corresponding interception point *ServerPreInvoke*.
- When the servant completes the processing of the requested operation, the ORB at the target side gains the control of the invocation processing. At that point, we identified the corresponding interception point *ServerPostInvoke*.
- The target side ORB communicates the reply for the requested operation back to the client side ORB, which in turn receives that reply and hands it back to

the client. Before the client receives the reply, we identified the corresponding interception point *ClientPostInvoke*.

- During the processing of a particular request, at both client and target side ORBs, system exceptions may be thrown. Because this can happen at any activity an ORB processes, we identified the corresponding interception point *SystemException*.

Several other interception points may be of great interest at request level, for example the ability to intercept the ORB's identification of a servant on the target side. To define the semantics of this activity and to include this in the specification, is an item for future work.

Network Related Interception

During the implicit binding for a non capsule-local interaction between a client and the target object, an ORB actively establishes connections or accepts incoming connection requests. Such network connection related activities include the choice of a transport endpoint to listen on at the server side, the choice of a transport endpoint for connection establishment and the actual connect procedure in the client ORB, the acceptance of an incoming connection and connection closure, performed by either the client or server side ORB.

Network level interceptors may be involved during such activities at five interception points related to connection management activities:

- The *PreAccept* interception point relates to the choice of a transport endpoint to listen for incoming connection requests at the server side,
- The *PreConnect* interception point relates to the choice of a transport endpoint for connection establishment, just before the actual connection establishment, at the client side - the object reference of the target object is known, which contains a number of transport protocol specific profiles;
- The *PostAccept* interception point relates to the acceptance of an incoming connection request at the server side - the selected and activated transport endpoint for a particular network connection is known.
- The *PostConnect* interception point corresponds to the activities done by a client side ORB just after successful or unsuccessful connection establishment - the selected and activated transport endpoint for a particular network connection is known.
- The *Close* interception point relates to activities to close a network connection, performed at either a client or server side ORB.

The network connection interception concepts can be applied to connection oriented networks. Although the use of TCP/IP in the CORBA world is common, more work is needed to develop concepts for interception of network related activities for the usage of connectionless network. One use of a connectionless network is already proposed by the GIOP over SCCP mapping in [1][8], another one for GIOP over IP-M and UDP has been presented in [3].

Management of Interceptors

Interceptors can access and modify information of object interactions, passed within the ORB core. In this way, they work together with the ORB in a very tight manner. Nevertheless they are entities, which should be distinguished clearly from the ORB core.

The lifetime of an interceptor seen from an information viewpoint is only correlated with the ORB i.e. an interceptor cannot exist without the ORB it belongs to. During the lifetime of the ORB, new interceptors can be created and existing interceptors can be deleted. Depending on the lifetime of an interceptor, the visibility of an interceptor to the ORB could be defined such that a certain interceptor may or may not take part in an invocation.

Interceptors are not assumed to inform the ORB of any modification they make to the information they receive before or after an invocation. The ORB has to trust an interceptor not to harm to the whole system in terms of security, functionality and error recovery. Furthermore no standardization of interceptor functionality can be done, so that ordering constraints on the invocation of different interceptors of the same interceptor type have to be applied from another entity.

This leads to the idea of an interceptor registry, which is responsible for registering and de-registering interceptors and for applying invocation constraints on the registered interceptors. The interceptor registry is the only entity that controls which interceptors to call during each invocation. In this role it has no knowledge about the functionality of each interceptor in principle, but classifies each interceptor according to the different interceptor types. Using this information, the interceptor registry can instruct the ORB to call the appropriate interceptors at each stage of the invocation.

The classification of an interceptor is done upon registration by providing the appropriate information about the nature of the interceptor. A registered interceptor is then able to intercept subsequent invocations. The question of whether a registered interceptor should be used in invocations or not is controlled by the interceptor registry on behalf of the application. No decisions are made by the interceptor registry itself, since the only appropriate information about the interceptor it has, is the interceptor type.

For the same reason that the interceptor registry cannot decide for itself whether to enable or disable a certain interceptor during invocations, it cannot make decisions about the ordering of interceptors, i.e. which interceptor to call first and which to call last during an invocation. Again this decision is up to the application, which must have appropriate knowledge about the registered interceptors. The interceptor registry assists the application through providing appropriate information about the current invocation order of the registered interceptors and providing a mechanism for reordering, but has no responsibility for defining the correct ordering of the interceptors.

To correlate an incoming reply with a previous outgoing request, interceptors often need to store some information about certain invocations for later evaluation. This requires some kind of cookie, which can be filled with information and can be accessed by the interceptor. These cookies, held within the interceptor registry for each registered interceptor, are created during the invocation phase of the interceptor and are made available to the interceptor again during the response phase.

Computational Model

After completing the phase of defining interception points and their related information, we developed a computational model for the interactions between the ORB runtime system and interceptor objects as well as computational concepts for the management of interceptors. The purpose of the computational model for interceptors is to define the interfaces and operations which are provided by both the ORB runtime system and the interceptors. Beyond that, the architectural model for interceptor invocations, the related interceptor management concepts and instantiation semantics are specified. A main goal was to ensure the portability of interceptor implementations across different ORB products.

More than one interceptor may be registered with an ORB for a particular interception point at the same time. These registered interceptors are invoked during the ORB's processing of an invocation.

We found three basic architectures for how an ORB may invoke interceptors:
- Daisy-chained invocation - The ORB calls the first interceptor, which in turn calls the second directly or indirectly, and so on.
- Serial invocation - The ORB calls each interceptor at each interception point, and the invoked interceptor returns control to the ORB before the ORB invokes the next interceptor.
- Conditionally serialized - The ORB calls each interceptor serially, but may omit an interceptor based on some constraint or condition.

A main advantage of daisy chained interceptor invocations is stacking of interceptors. In daisy-chained invocations of interceptors, there is no need for the ORB to explicitly remember, which interceptor was invoked during the request processing and should therefore be invoked again during reply processing. On the other hand, daisy-chained interceptors require different request related operations to handle synchronous, deferred synchronous and asynchronous requests. Moreover, an application of daisy-chained interceptors together with time independent invocations in CORBA Messaging [11] seems to require another set of operations to handle requests. Finally, a serial native ORB architecture would have to jump through hoops to emulate a daisy-chained architecture.

Serialized interceptor invocations require explicit mechanisms within the ORB to register which interceptors were called while request processing in order to invoke those interceptors again during reply processing in reverse order. On the other hand, this more complex ORB behaviour will be compensated by a number of advantages. First, a serial architecture for portable interceptors can be implemented over a wider variety of existing ORB implementations. A serial architecture is less sensitive to the various types of invocations that may be made in a CORBA client, including those invocation types supported by CORBA Messaging. Due to the mentioned advantages of serialized interceptor invocations and with respect to ease of implementation using existing ORB sources and the proprietary filter technologies of some ORB products, we decided to base all interceptor specifications on the serialized approach.

Another design consideration regards the number of interceptor local objects involved in the processing of a request or reply at an interception point. Each interceptor local object fulfils a particular functionality, e.g. an interceptor local object that implements a transaction service, or another that supports a certain management task. The question now is, whether such an interceptor local object should be instantiated each time that interception point is reached. Interceptor instantiation can be managed in two ways. An interceptor instance is created for each interception processing, e.g. an instance creation for each request/reply or an interceptor instance is created once and will be activated for each interception processing, e.g. for each request/reply. The first approach is more flexible, but requires extensive management capabilities as well as standardized factory interfaces for interceptors. The second approach is not as flexible but doesn't require such extensive management capabilities as the dynamic one.

Another criterion for interceptor processing is whether an interceptor may affect the ORB's behaviour with respect to a certain invocation processing or not. If an interceptor is not allowed to change the way an ORB controls the request processing, this interceptor can only be applied to monitoring tasks. If an interceptor affects the ORB's interaction processing, this may be done in several ways. A request related interceptor may change parameters of a request or reply message, but not the control flow with respect to the handled interaction. Changed request or reply parameters may or may not include changes to the service context of an interaction. A request related interceptor may change the control flow of an ORB regarding a particular interaction. This means for example, that a request related interceptor registered at the client side with respect to a certain object may produce a reply for a given request and send this reply back to the client directly. A network level interceptor may or may not send a given byte stream by itself, i.e. the interceptor instead of the ORB core sends the byte stream through a network API.

To identify to what extent an interceptor influences the ORB's behaviour with respect to a certain interaction processing, we defined capability sets for interceptors support within a particular CORBA runtime environment. These capability sets include

- Monitoring capability - an interceptor is only able to monitor the ORB's processing of a certain interaction,
- Parameter changing capability - an interceptor is able to change parameters of a certain interaction, like operation request parameters or service contexts,
- Control flow changing capability - an interceptor is able to change the ORBs normal control flow while processing a certain interaction.

3.2 COM/DCOM Interception Mechanisms

The architectural layering of COM is different from CORBA, and the interlocking of the COM runtime system with the Windows system is very tight. For this reason, it is very difficult to extract information from a (D)COM method invocation in an uniform way.

The extraction of method call information can be done on the component side using a message filter. A message filter is a component that exposes the interface IMessageFilter. Through this interface, the COM runtime system tells about incoming calls to components. The method that will be called in this case, is:

```
DWORD HandleInComingCall (
[in] DWORD dwCallType,
[in] HTASK htaskCaller,
[in] DWORD dwTickCount,
[in] LPINTERFACEINFO lpInterfaceInfo
);
```

A message filter component can extract the interface identifier from the iid parameter, and the method number. The method number is actually the offset of the called method in the virtual function table of the object inside the component implementing the called interface. On this point at least the interface and the invoked method can be extracted. In contrast to CORBA there is no easy way to extract the method name or method parameters for management purposes.

This could be overcome with COM+, the successor to (D)COM. COM+ (version 1.0) introduces interceptors for attaching standard services (e.g. MTS) to components. Microsoft promises for future version of COM+ the introduction of user-defined interceptors. It is supposed that these user-defined interceptors can play a similar crucial role in management, like interceptors in CORBA.

4   Management Service for CORBA Based Middleware Platforms

The Portable Interceptors framework, as the base for management information collection and management policy verification, was implemented during the project in order to verify its applicability for management purposes. In general, the interceptors specification was used for interactions between the specific management interceptor implementations and the ORB core, while the management interceptors interact with the environment using specific management channels (see Figure 3).

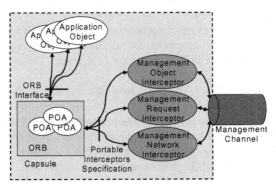

Fig. 3.   Management Interceptors Connected to the ORB and Management Channels

In more detail, a management specific interceptor object interacts with the ORB in a portable way to obtain object interaction information as well as to set and verify certain management policies. On the other hand, such interceptors interact with management channels to notify management-relevant events which occur during an ORB's object interaction processing. Furthermore, management interceptors receive management policies from such management channels. In that sense, management interceptors are context specific software components that interact with the platform core components in a portable, product independent manner.

The implementation of the portable interceptors framework focused on commercially available ORB products, we chose to use Inprise' VisiBroker, IONA's OrbixWeb and OOC's ORBacus. For these products, different implementation strategies were applied. While for VisiBroker and OrbixWeb, the product specific proprietary filtering mechanisms were wrapped in order to comply with the specification, for the ORBacus, the available source code was extended in order to support the interception of object interaction events presented above. Our management interceptors can be plugged into the interceptor framework implementation for each product in a portable way.

Management interceptors interact on one side with platform core components, like Object Request Brokers. On the other side, they communicate management-related notifications to management applications and receive management policies from those. For that purpose, the document type definition for management-related events and policies is mapped to CORBA specific constructs for transmission purposes. To achieve this, we defined a generic mapping between XML DTD constructs and constructs of the Interface Definition Language (IDL) of CORBA. This mapping in general uses IDL valuetypes as target constructs for XML elements. Attributes of elements are mapped to members of their representing valuetypes. This language mapping was applied to both the management-relevant events specification and the management policies definition, resulting in IDL definitions that represent the XML specification. These IDL definitions are used as communication elements between a management interceptor and management applications via management channels. Management channels are realized using the Notification service adopted by OMG [12]. Each conceptual management channel maps to a notification channel within a CORBA platform. The complete picture of how a management interceptor collects management information is depicted in Figure 4.

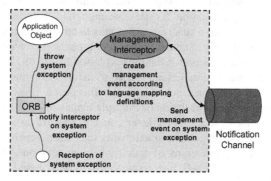

Fig. 4.   Collection of Management Information

If for instance an ORB receives a system exception as result of a operation invocation by an application object, it first verifies, whether an interceptor is registered for the SystemException interception point. If so, then the registered interceptor is notified on the happened system exception. The management interceptor in turn creates a IDL construct, that represent the according management event and sends that event through a notification channel to a management application. The application object, that originally invoked the operation that resulted in the system exception, receives that exception as usual.

If on the other hand a management policy is received by the management interceptor, it uses the standard ORB interface to override policies for the current thread of invocation. The mapping between management policies arriving through the management channel and those policies that can be set at the ORB is currently implemented as known by the interceptor object.

5   Conclusions and Further Work

Management services for middleware platforms must be provided with hooks within the core platform components so that basic object interaction processing can be monitored and controlled. "Management transparency" is an extremely desirable feature of heterogeneous systems - it should be possible to achieve end-to-end management in a distributed system which is built using a range of different middleware products and technologies. The use of Interceptors in CORBA and COM+, as described in this paper, offers the prospect of being able to "break out" of a specific middleware technology when required, which is an essential capability if portable management services are to be realized. In addition, some degree of standardization in the mechanisms for extracting relevant state information (i.e. Events) and inserting control information (i.e. Policies) is required. The correspondence or relationships between the idioms used in particular technology domains must be established where it is necessary to exchange management information between domains.

It is not envisaged that all middleware management will be restricted to some common subset of concepts applicable to all platforms. In other words, not all management services will be portable. Particular products are likely to offer management functionality that is product-specific. Equally, families of products (e.g. CORBA ORBs) are likely to have some management functionality in common. Beyond this, there will be a core of management information which is applicable across technologies. In addition a neutral reference representation and syntax must be chosen. XML seems to be a suitable choice for a standard representation. It is flexible enough to represent a very wide range of structured information, has a strict enough syntax to allow automated processing and has broad industry acceptance. Standard document types will be required to define the structure of management information but the text-based nature of XML documents and their straightforward structure makes automated transformation realistic.

There are several open issues, not addressed in this paper. These include techniques and concepts to support the correlation of multiple management events in heterogeneous distributed systems. In addition, while the concepts seem to provide many attractive features for introducing management functionality into heterogeneous middleware systems, the performance implications of specific implementations, based on CORBA Portable Interceptors or on COM+ interceptors need to be assessed.

References

[1]   Fischbeck, Kath: *CORBA Interworking over SS7*, in Proc. of ISN'99
[2]   Fischbeck, Holz, Kath, Vogel: *Flexible Support of ORB Interoperability*, in Proc. of Interworking ´98
[3]   Halteren, Noutash, Nieuwenhuis, Wegdam: *Extending CORBA with specialised Protocols for QoS Provisioning*, in Proc. of DOA'99
[4]   W3C Rec.: *Extensible Markup Language V. 1.0*
[5]   ITU Rec. X.901-X.904: *Reference Model for Open Distributed Processing*, ITU-T '95
[6]   OMG: *The Common Object Request Broker Architecture: Architecture and Specification*, Revision 2.3, OMG docs. formal/99-07-01 to formal/99-07-28
[7]   Expersoft et.al.: *Portable Interceptors Joint Initial Submission*; OMG doc. orbos/99-04-10
[8]   AT&T et.al.: *Interworking Between CORBA and TC Systems*, OMG doc telecom/98-10-03
[9]   OMG: *Portable Interceptors Request for Proposals*, OMG doc. orbos/98-09-11
[10]   BEA Systems et al: *Portable Interceptors*, OMG doc. orbos/99-12-02
[11]   OMG: *CORBA Messaging*, OMG doc. ptc/00-02-05
[12]   OMG: *Notification Service Specification*, OMG doc. telecom/98-11-01

Meta-management of Dynamic Distributed Network Managers (MEMAD)

Ran Giladi[1,2] and Merav Gat[2]

[1]InfoCyclone Inc.
ran@infocyclone.com
[2]Communication Systems Engineering Department
Ben-Gurion University of the Negev, Beer-Sheva 89105, Israel
ran@bgumail.bgu.ac.il

Abstract. Distributed network management systems (NMS) have become a crucial necessity, especially for overcoming centralized NMS restrictions such as scalability and inefficient use of network resources. Such systems will also be instrumental in meeting the need for high-power computers and storage capabilities on the NMS platform. Modern technologies used in distributed NMS include management by delegating agents and mobile codes. These methods lead to the creation of a hierarchical architecture, since it simplifies management of the distributed agents. Peer management results in a dynamic, survivable and efficient way of managing the network, but it requires a complicated metamanagement mechanism to handle the managers. This study suggests an architecture for this purpose. We term this model Meta-Management of dynamic Distributed network managers (MEMAD). The purpose of MEMAD is to enable Peered Distributed Managers (PDMs) to manage the network by executing delegated or predetermined common management tasks. MEMAD defines a small, shared, replicated, and partitioned database as well as inter-communication SNMP based primitives for providing PDMs with the ability to cooperate efficiently in managing the network.

1   Introduction

Most network management systems available today are based on a centralized network management architecture. This simple architecture is composed of two entities: the Network Element (NE) which contains an Agent (a server with a small footprint), that holds management information concerning its managed NE, and a Manager (a client that runs applications). The Manager provides the management applications with a network view while it manages the agents by polling and processing their information. The agent and the manager communicate by using the standard simple network management protocol (SNMP) [1]. The manager performs most of the management work – polling, processing and controlling the Network Elements (NEs), while the agents access the NE information base in order to retrieve or update management information.

A. Ambler, S.B. Calo, and G. Kar (Eds.): DSOM 2000, LNCS 1960, pp. 119 - 131, 2000.
© Springer-Verlag Berlin Heidelberg 2000

In recent times, Web-based management systems have begun to use a technology that somewhat differs from SNMP systems in that the management station communicates with its users or applications via Web-based GUI (e.g., HTML, XML) [13,15]. Web-based technology also allows Network Elements to respond with HTML pages instead of with SNMP, CMIP, or DMI messages to the management station [14].

In large and complex networks, the managers have to control a vast number of agents and perform all the information processing to the extent that the managers are loaded with information coming from all the agents. As a result, the managers' limited resources and network accesses bottleneck the management system. Additional problems such as increased network congestion, a high probability of disconnecting agents from the manager, slow responses from managers, and lack of redundancy have thwarted the centralized network management architecture from efficiently managing today's networks.

Of late, we have begun to witness a new approach to network management that, in principle, decentralizes network management, or better yet distributes it. The distributed network management approach deals mostly with the distribution of network management applications by delegating management tasks to other management entities for execution. Modern technologies presently used in this kind of distributed network management are usually hierarchically structured, since management of the distributed agents is simple and inherent in such structures.

Management by delegation (MbD) was among the first mechanisms to be proposed for hierarchical network management [2]. By delegating management tasks to specific agents, the manager's load is decreased and the tasks are executed close to the Network Elements (NE). This leads to a decrease in management traffic, higher availability of the NEs, and less dependency on the manager. However, for executing management tasks delegated by a manager, it has become necessary to devise a flexible and generic mechanism to serve as an extension to the agent.

Another approach to hierarchical network management [3] introduced the SubManager as a dual function entity. This manager delegates management tasks to the SubManager with access authorization to the relevant NEs and their agents. The SubManager then executes the task while polling the agents for the required information. This hierarchical model uses an SNMP framework and extends the Management by Delegation model by enlarging the scope of management of the dynamically extended entity (the SubManager).

Other approaches to distributed network management that have been suggested include the addition of RMON (remote monitoring) capabilities [4] to the executing entity [5], or the use of a Mid-Level-Manager and SNMP protocols. In the latter, the Mid-Level-Manger is provided with a new MIB (management information base) definition [6], which makes it possible to receive delegated management scripts and execute them on the NE agents, as well as other distributed management tasks.

Recent technologies, such as mobile code, have resulted in the Mobile Agent [7] approach. This technology implements the delegation mechanism in a slightly different manner, in that in order to achieve better network resource utilization, every management task is delegated and executed by several agents and servers. The use of recent technologies, such as mobile code, intelligent agents, etc. [8-10], provides more sophisticated and efficient distributed methods than the various hierarchical methods commonly implemented (for example, peer distributed network management and fully distributed network management [11] as well as improved delegation mechanisms [12]).

Nevertheless, most of the afore-mentioned approaches lack a proper mechanism for controlling the distributed managers, especially in non-hierarchical distributed management systems and their derivatives. We call this problem the metamanagement problem because while the execution burden is shifted from the centralized manager to other mid-level management entities, the (human) manager still has to plan, configure and control the participating mid-level managers, the task distribution, and the execution process. These activities are essential for preventing redundancy, contradiction, and inconsistent management activities. For large scaled networks this task is virtually impossible if not automated.

In this work, we suggest a peered distributed network management model, which is based on a meta-management distributed architecture (MEMAD) that handles Peered Distributed Managers (PDMs). In this system, each PDM can implement a fully functional management application and execute its own management tasks. The PDM can also receive and execute a delegated task or a mobile agent from some management application.

In every distribution system, we are confronted with inherent difficulties and overhead regarding the management and the synchronization of the distributed elements. The suggested MEMAD architecture enables the PDMs to operate optimally and dynamically according to the network, to the PDMs states and to the applications, while utilizing minimum network resources. MEMAD deals with the metamanagement of the PDMs, rather than the managerial tasks the PDMs have to perform. Self-configuration, information and task distribution, backup, recovery, remote, and proxy operations of the PDM will be supported by MEMAD, regardless of the management activity the PDM is involved with for the management application. MEMAD does not deal with the distribution (by delegation, mobile agents or otherwise) of management applications; rather it guarantee that the management of the PDMs is carried out precisely, effectively, and efficiently.

Each PDM that uses MEMAD is an autonomous system that controls its own physical management domain. It communicates and cooperates with other PDMs, it responds to messages from management applications and PDMs, it learns, and it is proactive. The union of these physical management domains and the resulting response from the PDMs cover the entire network and the entire range of the NMS functions.

The organization of the paper is as follows: The following section presents MEMAD's architecture and the PDM model. The main algorithms are presented in section 3. Implementation and future work are described in section 4 and section 5 concludes the paper.

2   The MEMAD Architecture

The MEMAD architecture for peered distribution of network management deals with how PDMs cooperate to carry out network management tasks, regardless of whether the tasks are predetermined or delegated according to need. Each PDM manages its Network Elements (e.g., via SNMP) and is responsible for doing so in some territory that is defined in terms of space or function. MEMAD ensures that all PDMs are synchronized and cover the desired network efficiently and without overlapping or holes. Although we implemented MEMAD to deal with network scope in terms of space, it can be implemented also in terms of functions.

2.1   Objectives

MEMAD was designed so that all legacy systems can be integrated with a consistent peered or fully distributed network management. In other words, it allows SNMP agents to participate in a scalable and efficient distributed management system without altering the agents. MEMAD's components and primitives are kept as simple as possible, thus enabling all kinds of distribution mechanisms (peered or otherwise, that are using mobile agents or delegated tasks) to function optimally (optimal in the sense of network traffic, response time, scalability, processing power, storage capabilities, etc.). MEMAD enables distributed network management to be implemented without tedious human planning and configuration, while maintaining load balancing and moderate network resource usage.

2.2   Components and Network Setup

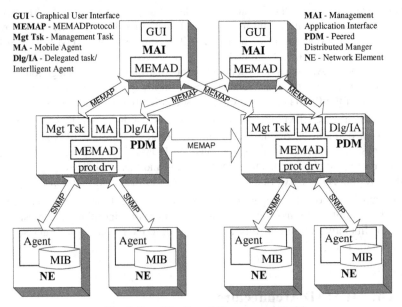

Fig. 1. Peer Distributed Management Entities

MEMAD defines a minimal set of entities as well as their capabilities (see Figure 1). They are described as follows:

- The *Management Application Interface (MAI)* is a very "thin client" which is separate from the manager and is used as an interface to the outside world. Through the management application interface and its GUI, a human manager can retrieve information and observe, hence, control the network.
- The *Peered Distributed Manager (PDM)* is the main entity of the model. This manager executes either a full management application or a delegated task on a limited group of Network Elements. The management application consists of a fixed set of management rules and tasks that enable the manager to perform independent activities on its Network Elements.
- The *Agent* remains within its traditional "thin server" architecture.

The PDM operates within the network where it manages the Network Elements (agents) for which it is responsible, and communicates with its peers and with the MAI addressing him, according to MEMAD definitions, as illustrated in Figure 2:

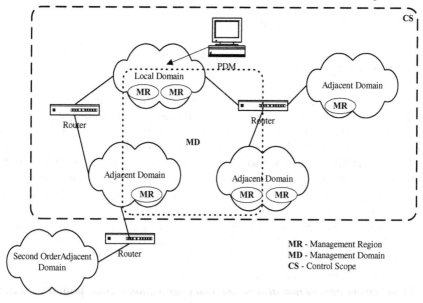

Fig. 2. Management Scope Definitions

- The *Domain* defines a sub-network (e.g., class B or C in IP networks). A *Local domain* of a PDM is the domain where this PDM is physically placed. An *Adjacent domain* of a PDM is a domain separated from the local domain by one router.
- The *Management region (MR)* is the collection of consecutive IP addresses within a specific domain. The *management region* of a PDM contains the agents to be managed by this PDM.

- The *Control scope (CS)* is the union of those domains in which a PDM can have any *management region*. In our work we restrict the control scope of a PDM to its *local domain* and all its *adjacent domains*.
- The *Management domain (MD)* of the PDM is its collection of *management regions* inside its control scope.
- The *Management weight (MW)* of a PDM is measured as the number of all the agents it manages.
- The *Local PDM* of a domain is the PDM that is physically placed in that domain, and the *Adjacent PDM* of a domain is the PDM placed in one of the adjacent domains.

We emphasize that the PDM has the sole responsibility for its management domain (unless it allows some other PDM to access its territory). However, several PDMs might reside in a sub-network (one domain). In that case, they will have to split the domain between them as equally as possible. There is also a situation when there is no PDM in a domain, and a PDM from another domain (usually an adjacent domain) "takes over" this domain as if it were its own. This "take over" can occur while the control scope is being reorganized, and will be carried out in as balanced a way as possible among all the PDMs in the system.

2.3   MEMAD Functions

In order to implement MEMAD we first had to construct a minimal set of functions. These functions include configuration, information distribution and aggregation (information regarding the network management itself), fault tolerance, and proxy management of various kinds, all of which are described below:

- The *Configuration* function enables a dynamic, adaptive and auto-configuration of the PDMs in the managed network, and includes primitives such as: "hello", "I heard you", "who manage", "I manage", "acquire", "my world", "manager down", "unmanaged MR" and "re-managed MR". Another set of primitives is concerned with load balancing of the PDM configuration and includes: "balance", "balance information", "balance start", "balance start information", "balance end", "update", "update others", and "update split".
- *Management information distribution and aggregation* include primitives such as: "query", "sub-query", "answer", and "sub-answer".
- *Fault tolerance* is a function that maintains a constant backup of the essential information and allows uninterrupted management and fast recovery. It includes choosing the right PDM to be the "hot backup" PDM, handling a partial replication of the metamanagement database in the network, and recovery processes. The extent of the replication is kept to a minimum so the network traffic stays low, while ensuring continuous operation and rapid recovery.
- *Proxy management* is a function that allows the PDM to share management tasks with other PDMs in the managed domain, or to allow other, remote PDMs to manage the domain. Primitives supporting this function include: "remote", "share", "data", "data OK", "lengthen", "lengthen OK", etc. These

primitives contain permission mechanisms that assure the security, period and scope of proxy management.

2.4   The PDM Database

Each PDM holds a management database that enables its management activities to be executed. The PDM has to manage its management domain in cooperation with other PDMs in the system at the same time that it performs its own distributed network management tasks. The PDM's database can be implemented using MIB tables (as we did in this study) or it can use other implementations.

The database contains a full description of the management domain and the control scope of the PDM. A description of other PDMs and their domains are sufficiently detailed in the database so as to enable MEMAD to operate.

The database is composed of three categories of information:
- A Class hierarchy that describes all the entities in the PDM's environment (device, link, node, role, etc.).
- A Collection of all the entities currently active in the PDM's environment and their detailed attributes.
- Detailed management information about the PDM's control scope: management regions, the entities residing in these regions, other active peers, management domains and management states, etc.

When the PDM is initiated, its database contains only its class hierarchy. In the process of identifying other peer managers (even before executing the "load balancing" algorithm), the PDM polls them for the necessary information about their management domains. Once the PDM finishes the initial data acquisition process, it updates the database with the relevant changes that occurred in its environment.

2.5   MEMAD Protocol (MEMAP)

The MEMAD protocol (MEMAP) was devised to enable PDMs to communicate both with each other and with the MAIs. The MEMAP is applied between the MEMAD layers in each PDM or MAI participating in the distributed management. We implemented the database using an MIB structure, and used SNMP queries to tunnel the MEMAP, although other methods are possible (e.g., CORBA, WEBM, etc.).

The MEMAP consists of a set of messages that enable primitives such as those described in section 2.3 to be transferred.

2.6   MEMAD Procedures

MEMAD ensures that all its participating components are able to provide a coherent and consistent view of the network, and that they share the managerial tasks equally. For this, MEMAD entities must inter-communicate and operate in the following way:

A user (or management application) approaches one of the MAIs for some management task (get or set some attribute in the network, analyze or produce collective information, etc.). This MAI uses its MEMAD layer to reach a default PDM (usually one that was placed in the same domain), and hands over the management task to this PDM.

This PDM becomes the "organizer" PDM, and activates a process, which we call the Integrator process. This process uses its PDM MEMAD layer to find the relevant PDMs for executing the required management tasks by checking the required tasks (e.g., IP addresses) with the management domains of all the PDMs listed in its PDM's database. The organizer PDM then retrieves the information, sets attributes, or initiates management tasks in these PDMs via their MEMAD layers. The PDMs operate with their Network Elements via their common interface, SNMP, CMIP, DMI, etc. The responses are then collected by the organizer PDM, aggregated (see, for example, [13]), and sent to the querying MAI.

In order to locate the relevant PDMs for the various managerial tasks, the PDMs must be configured in a way that makes them recognizable to all MEMAD participants. This configuration is a dynamic process, and is carried out by the PDMs themselves. The PDMs divide the network among themselves by assigning parts of the network to each of them as their management domains. Whenever there is a change in a PDM's status (e.g., a disconnection, a failure, etc.), the managed Network Elements are regrouped and assigned again to the PDMs, so that the management weight of the PDMs is distributed equally. A description of the underlying algorithms is provided in section 3.1.

Initially, there are no PDMs present, and no management activities. Any PDM joining the system defines its management domain. The first PDM to appear takes over the whole network. Other new PDMs will check with the active PDMs and build their own group of Network Elements and management domains, thereby decreasing the management weight from their peers. This procedure of reorganization is carried out using a load-balancing algorithm that will be explained in section 3.2.

This self-configuration, i.e., the dynamic definition of the management domains, also takes place when a PDM fails (while its backing PDM initiates the reorganization procedure), or on a periodical basis initiated by a local PDM.

Other procedures support the PDMs in maintaining a fault tolerance system, ease of operation and other special requirements. These include backup and recovery processes, remote management, and shared management processes. The backup management enables every PDM to be backed-up by one of its peers (see section 3.3).

The remote management procedure allows a temporary connection between two PDMs to be established. It enables a PDM to take a respite from management activities without being declared as failed, and it allows a remote PDM to take control over its management domain. A PDM that needs time off finds another PDM that agrees to serve as its remote manager for a predetermined period of time. During the remote connection, the remote manager takes responsibility for all management activities in the management domain of its connected PDM. When the remote connection ends, the original PDM returns to manage its own (updated) management domain. Another motivation for this kind of proxy management results from the need of a PDM to watch some domain in a detailed resolution which is required by its management application. In this case, the remote manager initiates the procedure.

The sharing management procedure enables yet another type of temporary connection to be created whose purpose is to lighten the management's burden. This PDM looks for a peer that will agree to share its management domain for this period of time. During the shared session the volunteer PDM is responsible for all the management activities of the managed Network Elements that it has received from the initial PDM. Upon ending the connection, the initial PDM fully remanages its management domain.

3   MEMAD Algorithms

This section presents some of the main algorithms of the MEMAD architecture.

3.1   PDM Initiation

The initiated PDM's database is empty, thus the PDM is not aware of any other PDMs or agents in the system. Neither is he aware of the control scope nor of any other domain. There is no "super-manager" from which the new PDM can get information, nor is there a "super-manager" to introduce the initiated PDM to other PDMs. Consequently, the PDM has to search and learn the control scope by itself and then introduce itself to its peers within this control scope. During this "introduction" the PDM can seek information from other PDMs so as to fill his database. The PDM initiates a load balance algorithm after it learns the control scope, and then becomes an active PDM, that is equal to its peers.

To learn about its control scope, the initializing PDM performs a topology analysis. After the control scope is known, the PDM discovers other PDMs by broadcasting introduction messages to all domains within its control scope. Every PDM that receives an introduction message acknowledges it with information on the control scope and on other PDMs in its local or adjacent domains. In other words, the initializing PDM discovers another PDMs when it receives information regarding their network view. Both the initializing and the discovered PDMs are synchronized, and the initializing PDM synchronizes also with the PDMs he learnt about from the discovered PDMs. This method enables scalability, because it is not necessary to reach all PDMs while initializing, and due to the asynchronous way the PDMs are discovered and updated.

3.2   Load-Balance Algorithm

This algorithm deals with the load-balance between the participating PDMs and the distribution of management domains among the PDM within the control scope. Load balancing means distributing the load of management activities as evenly as possible between the managers, but it can be based on any optimal criteria. Examples might include traffic, complexity of management tasks, a PDM load or response time, quantity of Network Elements, physical distance from Network Elements, network resources or their utilization, significance of Network Elements, reliability or quality of links and subnetworks, or a combination of these criteria. We implemented a simple objective function that is based on the quantity of the managed Network Elements and their network distance from the PDM, according to which the load-balance was computed. Thus, balancing the load in our implementation means that there exists an equal quantity of managed Network Elements per PDM in the vicinity of the PDM.

The load-balance algorithm is used when a new PDM enters the system or when a PDM terminates (gracefully or not), or on a periodical basis.

The algorithm is executed under the following constraints:

- Only one PDM at a time, within a certain control scope, can execute the load-balance algorithm. For the execution of load-balance it is necessary that all communicating PDMs be synchronized (non-communicating PDMs are not relevant).
- A PDM will always first be assigned to its local domain. It might also be assigned to other domains in its control scope, if they don't have any local PDMs.
- Management domains cannot overlap.

The load-balance algorithm consists of two consecutive stages: balancing the management load in the local domain, and balancing the management load in other adjacent domains that contain no local PDMs.

These two stages of the load-balance algorithms are almost identical and the only reason for the different balancing processes is the network distance criteria. According to the objective function, a PDM should manage the closest Network Elements it can locate. For this reason, the PDM should first balance the load in its own local domain and only then balance the load in the other domains.

To fully synchronize the load-balance algorithm, the PDM that executes the load-balance algorithm gets updated information from all its peers. Each PDM delivers information about its management regions, from which the balancing PDM can derive the management weight (MW) and the average distance attributes of the sending PDM. Additional data assembled by the PDM that executes the load-balance algorithm includes:

- Its distance from each management domain.
- The distance ratio (DR) of each management region (MR).
  DR = distance of MR from its original (local) PDM / distance of the MR from the balancing PDM.
- OMW - Optimal Management Weight (MW) in the domain:
  OMW = MW / number of PDMs in the domain.

Once all the required information is available, the PDM that executes the load-balance algorithm simply attaches (logically) the MRs with the lowest DR to its MD, until its MW reaches the OMW point. Afterwards, it balances the other PDM's load, but with one difference in the algorithm, namely, there is no DR parameter. In this case, the balancing PDM will try to remove an MR from one PDM and assign another PDM to it if the MR remains intact. The second stage (balancing the adjacent domains) is performed exactly as done for balancing the other PDM load in the first stage.

The load-balance algorithm is terminated when the balancing PDM notifies so to all its peers, and when it informs them of all the changes in their management domain (MD). Once a PDM receives a message of a changed MD, its responsibility is to actuate these changes by sending a relevant message to the other PDMs. These PDMs become the new managers of its prior management regions (MRs).

The synchronization of executing the load-balance algorithm among the PDMs in the CS is achieved by using a "balancing queue" which is synchronized and maintained by all PDMs. This queue lists all those PDMs who are waiting to execute the load-balance algorithm. A PDM can execute its load-balance algorithm only when it reaches the top of the queue.

3.3   Backup Management

The backup algorithm is required in order to increase MEMAD's reliability. It dictates that every PDM has another PDM backing it up. When a new PDM gets its MD it has to look for a PDM to back it up. The most suitable PDM is chosen according to a set of parameters: the shortest distance between the PDMs, the lowest MW, and the absence of any other backup responsibilities. The new PDM polls all its peers for the relevant information and only when it finds the right candidate does it initiate the backup request.

Once the backup connection is made, the backed-up PDM updates its backup PDM on a periodical basis. These update messages contain the necessary data to enable the backed-up PDM to recover and reenter the system without starting the full process of an initiating PDM.

If the periodical update is not received on time, or in the case when a "manager down" message is received from another PDM, regarding the backed-up PDM, the backup PDM will know that the backed-up PDM is faulty. The backup PDM will then try to establish a connection with its backed-up PDM, and if it fails, the backup PDM takes full responsibility for the backed-up PDM's MD.

4   Implementation and Future Work

We are currently implementing MEMAD on 30 small sub-networks, using Java and JMAPI. Screening the networks, learning its topology and locating SNMP agents is performed with Win-Socket32, while communicating with the SNMP agents is done with JMAPI. The communication between the PDMs is carried out using Java-RMI. A combination of an SNMP-like protocol and RMI was created for maintaining an SNMP-based MIB to serve as the PDM's database.

The management task we tested dealt with the auto-discovery of network topology, which is carried out by several PDMs in parallel. We intend to extend this implementation for executing vast accounting management applications that call for continuous sampling of equipment in various places of the network, storage of huge amounts of data, and analysis of aggregated information.

5   Conclusions

This work concentrated on the metamanagement problem of distributed network management systems (MEMAD). We described the essential features of such systems and designed and implemented the necessary algorithms for carrying out these features. Finally, the system was tested by an application for network auto-discovery.

The results showed that the distributed system configured itself very quickly, and was able to adjust itself to all changes in PDM availability, network fluctuations, and management application requirements. MEMAD was implemented for minimum distributed network management tasks, allowing all kinds of distributed mechanism to be employed. MEMAD can be implemented by any distributed network mechanism (e.g., hierarchical, delegated management tasks, intelligent agents, mobile codes, etc.). Finally, MEMAD exhibited a dynamic auto-configuration, fault-tolerance and responsiveness to management tasks.

Although we implemented MEMAD to deal with network scope in terms of space, it can be implemented in terms of functions as well.

References

1.   Subramanian, M.: Network Management: Principles and Practice, Addison-Wesley, Reading, MA (2000)
2.   Goldszmidt, G., Yemini, Y.: Distributed Management by Delegation. In: the 15th International Conference on Distributed Computing Systems, Vancouver, British Columbia (1995)
3.   Siegl M., Trausmuth G.: Hierarchical Network Management: a Concept and its Prototype in SNMPv2. Computer Networks and ISDN Systems **128** (1996) 441-452
4.   Waldbusser S.: Remote Network Monitoring Management Information Base. RFC 1757 (1995)
5.   Williamson B., Farrell C.: Distributed Management using Remote Monitoring Management Information Base (RMON MIB) and Extensible Agents. Technical Report, Curtin University of Technology (1996)
6.   IETF Distributed Management Working Group (disman), RFC2591-2 and various drafts (http://www.ietf.org/html.charters/disman-charter.html) (2000)
7.   Baldi, M., Gai, S., Picco, G.: Exploiting Code Mobility in Decentralized and Flexible Network Management. In Proceedings of the 1st International Workshop on Mobile Agents (MA'97), Berlin, Germany (1997)
8.   Cheikhrouhou, M., Conti, P., Labetoulle, J., Marcus, K.: Intelligent Agents for Network Management: a Fault Detection Experiment. In Proceedings of the 6th IFIP/IEEE International Symposium on Integrated Network Management (IM'99), Boston, MA (1999) 595-609
9.   El-Darieby, M., Bieszczad, A.: Intelligent Mobile Agents: Towards Network Fault Management Automation. In Proceedings of the 6th IFIP/IEEE International Symposium on Integrated Network Management (IM'99), Boston, MA (1999) 611-622

10. Zapf, M., Herrmann, K., Geihs, K.: Decentralized SNMP Management with Mobile Agents. In Proceedings of the 6th IFIP/IEEE International Symposium on Integrated Network Management (IM'99), Boston, MA (1999) 623-635
11. Sahai, A., Morin, C.: Towards Distributed and Dynamic Network Management. In Proceedings of the IEEE/IFIP Network Operations and Management Symposium (NOMS'98), New Orleans, Louisiana (1998)
12. Breitgand, D., Dolev, D., Shaviner, G.: HAMSA: Highly Available Management System Architecture, Technical Report, Hebrew University of Jerusalem (1999)
13. Anerousis, N.: A Distributed Computing Environment for Building Scalable Management Services. In Proceedings of the 6th IFIP/IEEE International Symposium on Integrated Network Management (IM'99), Boston, MA (1999) 547-562
14. Martin-Flatin, J.P.: Push vs. Pull in Web-Based Network Management. In Proceedings of the 6th IFIP/IEEE International Symposium on Integrated Network Management (IM'99), Boston, MA (1999) 3-18
15. DTMF standards concerning DMI, WBEM, DEN, http://www.dmtf.org, 1998-2000.

Service Contracts Based on Workflow Modeling

Holger Schmidt

Munich Network Management Team
University of Munich, Dept. of CS
Oettingenstr. 67, 80538 Munich, Germany
Phone: +49 89 2178 2165, Fax: +49 89 2178 2262
schmidt@informatik.uni-muenchen.de

Abstract. The increasing importance of IT infrastructures and the complexity of IT services in each company often results in endeavors to outsource IT services. This implies meeting the challenge of service quality. Therefore, it is important to sign a service contract for the outsourcing partnership supporting effective usage and management of the service. It must be possible to monitor and manage the service in a constructive and fast manner.
This paper presents a customer-oriented approach for specifying service contracts. The idea is to combine service contracts with workflow concepts. The knowledge about design and management of workflows can be used to specify a service contract constructively supporting operation and usage of complex services. The use of the customer's business processes as a basis for the contract ensures a customer-oriented service view. Workflow concepts allow both, specifying non-ambiguous contracts and constructive instructions for usage and management of services by the customer. Combined with a suitable contract structure a controlled dynamics for the service contract is possible.
Keywords: Service Contract, Service Level Agreements, Workflow, Business Process, Outsourcing.

1 Introduction

The increasing dependence of a companies success on IT infrastructures requires to face the challenges of service quality. The complexity of the needed IT services is growing. Therefore, the design and operation of services is often outsourced to independent internal departments or to external partners.

To specify the requested service customer and provider sign a service contract. Such contracts define the functionality of the service and the required service level. The contract is an important source of information for the provider's service management because the service provider usually does not know much about the scenario the service is used in. Thus, the quality of the contract is one important parameter for the success of outsourcing relationships.

Keeping the service quality on a reasonable level is the challenge of modern services. It is also a task of service contracts to ensure that the customer has sufficient management facilities to fulfill the customer's duties, to customize the service according to the users' needs and to control the provider's service provisioning. Therefore, management interactions need to be specified in addition to service level and functionality.

A. Ambler, S.B. Calo, and G. Kar (Eds.): DSOM 2000, LNCS 1960, pp. 132–145, 2000.

In this paper the term management means the service management crossing the domain boundary from the customer domain to the provider domain. Management of the service implementation done by the provider is considered as operation.

The specification of interactions is difficult if it is based on a relatively informal collection of rules and statements. The temporal aspect of interactions cannot be represented adequately and it is hard to master the complexity with unstructured methods. From such methods many contracts for today's IT services result. Interactions can be modeled as a process. Production processes or business processes are described by workflow concepts. Therefore, this paper proposes the usage of workflow concepts for designing and writing of high quality service contracts for IT services.

For demonstration purposes a scenario is presented in section 2. Sections 3 defines the term service contract, derives requirements and identifies the contract elements. A graphical representation of workflows is introduced in section 4. Section 5 presents our approach starting with the contract structure followed by the process model and a classification of processes. Section 6 shows a simple process for contract design and its application. Finally, our approach is discussed in section 7. The last section draws a conclusion and introduces future work.

2   Scenario

To clarify the presentation of our approach and for demonstrating the potential complexity and investments of today's services a scenario is developed. (see figure 1). The scenario originates from experiences in several projects with industrial partners. It shows the complexity and high investments a custom–designed IT service can require today.

A company selling its high priced customizable products over an international network of dealers wants to connect those with its own central IT infrastructure. This permits the dealers to customize and order products online. Additionally, multimedia information material is available online for product presentations. Furthermore, up to date information on the stocking, production date, date

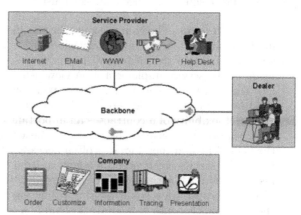

Fig. 1. Scenario

of delivery, etc., is available for sales conversation of the dealers with their customers. Each product is individually traceable during the production and delivery processes. Similar facilities are available for after–sales contacts. The applications supporting this functionality are hosted by the company itself.

The company has a contract with a service provider for the functionality and necessary management facilities described next. The dealers can use Internet connectivity as additional features with special support for email, WWW and FTP. All communication between dealers and the company is encrypted for security reasons. A firewall secures the Internet gateway. The infrastructure is IP based which requires the operation of all supporting services like management of domain names and IP addresses including the necessary infrastructure. The dealer's connectivity to the provider's backbone is realized by leased lines or ISDN using components at dealers locations owned by the service provider. Finally, the usually quite unexperienced IT users at the dealer locations need a help desk supporting them in day to day connectivity problems. The service provider insists on a long–term contract of several years because the provider wants to amortize the investments in design, hardware, software and staff.

3   Service Contract

This section defines service contract and identifies requirements as well as its elements. A *service contract* is an agreement between customer and provider specifying the service functionality and all management interactions of the customer–provider relationship including the required service level for both.

3.1   Requirements

Common requirements like validity, completeness, consistency or unambiguity must be fulfilled by all types of contracts. But there are additional requirements for service contracts that specify complex IT services in a long–term customer–provider relationship. A service contract must be:

Customer–Centric: The contract must be specified in a terminology originating from the customer's usage scenario. The service implementation should be transparent to the customer. The service specification may not include terms of the service implementation view usually preferred by the provider. But this must not lead to restricted expressiveness. All necessary details must be expressible while allowing the abstraction of well known facts.

Dynamic: Flexibility of a contract is an important factor for a successful long–term relationship because the needs of a customer change relatively fast in the field of IT and therefore it is often necessary to adjust some parameters. Especially service levels need an adaption from time to time to fit customer needs. This must be possible in a controlled way. Not all parameters are tunable because they could require a different implementation. Therefore, the dynamic elements in a contract should be selected.

Constructive: Management facilities for the customer are an essential element in a contract that allows the customer to monitor and influence the provider's service operation to a certain degree, e.g., to customize the service to the users' needs. The management interactions should be used to actively prevent or at least constructively solve problems in cooperation. Operators as well as users should be guided to become active by a service contract if the situation requires intervention. The contract must be understandable for both, users and operators. Therefore, a contract requires concrete and workable instructions which are supportive for usage and management. All other management interactions like ordering, accounting, maintenance, etc. need to be explicitly specified, too. This way the service contract is present during operation and usage not just in a legal proceeding.

Summarizing, to reach these targets a constructive contract is needed which is customer–centric, service–oriented and controlled adaptable. Furthermore, it must be concrete, measurable, realistic, understandable and workable, but without loosing expressiveness. That means all contracting partners know their rights and their duties.

3.2   Elements

The rights and duties are specified by three groups of information: legal, organizational and technical information. To be considered valid a contract must comply with all formalities, e.g., names, addresses and signatures of the contracting parties are needed. This *legal information* is important for a contract, but it is regulated by law. Therefore, this paper focuses on organizational and technical aspects of the customer–provider relationship.

The service is described by *technical information* on functionality, capacity, quality and the interfaces for usage and management. Functionality describes what the service supports to do, not how it is done. Capacity and quality criteria specify the service level that must be fulfilled. The usage interface is the service access point. A management interface is needed for monitoring and influencing the service from the customer's domain.

The *organizational information* includes all interactions between customer domain and provider domain for the usage and management of the service. A usage interaction is for example ordering a product. Interactions like adding a new dealer or problem handling are management interactions.

3.3   Related Work

Most service contracts for complex services specify a set of rules and statements which are based particularly on experiences of provider and customer. This very informal procedure does not help in fulfilling the raised requirements. It is very difficult to write high quality contracts, if there is no structure or methodology guiding the design process which also allows to involve the service users.

An approach for specifying contracts in federated environments is presented in [1]. A service consists of several components. For each component quality metrics and interdependencies to other components can be specified. This straightforward approach allows to verify the service quality by evaluating relations of measured values provided by each component. The focus lies on the service implementation. This complicates the use for complex services because customers are not interested in the numerous measured values of components implementing the service. The implementation view is useful for providers but too difficult to understand for customers which are not familiar with the possible service implementations. Furthermore, this approach does not support management from the customer's domain.

Another approach defines a model of electronic services for long–term relationships between customer and provider [7]. Contracts are negotiated between customer and provider to guarantee the availability of a service for repeated use in an open service market where one–time service usage dominates. The approach is based on virtual resources which represent the interface between customer and provider. They are mapped to physical resources on service usage. This approach specifies the interface between customer and provider hiding the service implementation but still focusing solely on the service usage. Interactions not concerned with usage like problem handling are not considered.

There also exists some work of the TeleManagement Forum on processes needed to manage telecommunication services [10]. The main focus in not the creation of contracts but the automation of business processes. After completion this standardization effort can simplify the specification of contracts.

No work for service contract specification regarding the interactions which support usage and management of services from the customer's domain is known. The approach presented in this paper uses workflow concepts for service contracts to support usage and management of services from the customer's domain and to enable a systematic design of contracts.

4 Workflow Modeling

The most important standard body in the area of workflow technology is the Workflow Management Coalition (WfMC). This organization has defined a reference model [11] and terminology for workflow modeling [12]. Based on this work there exists a graphical representation for workflow models [8], [9], [4] which also allow the syntactical verification of the models. Because UML [6] is commonly accepted and the Object Management Group also specified the representation of workflow graphs [5] which is based on the work of the WfMC, UML should be used in this paper for workflow models. This is no restriction because the graphs can easily be mapped to the other representations.

For the representation of a workflow we use activity graphs (see figure 2) suggested in [5] in a simplified version compared to the activity graphs specified in [6].

Workflow graphs are named at the top left corner of the graph. It is possible to group several parts of the graph in so called *swimlanes* which are separated by thin vertical lines. This can be used for example to group activities to organizational units or to actors referenced by the name at the top of a swimlane.

Activities are represented by rectangles with rounded corners marked with the activity name. The model supports recursion by *composite activities*. Such activities are marked with a small symbol at the right side of the rectangle. A composite activity is modeled as a workflow elsewhere which is references by its name.

A workflow starts at a *start symbol*, a small filled circle. It finishes either at an *end symbol*, two circles while the inner one is filled, or at a point where no further action can be executed because all conditions allowing the execution of a path evaluate to false.

Conditions appear in the graph in two ways. They can be explicitly modeled by a condition symbol drawn as a rhombus with conditions in square brackets at each leaving transition arrow. A condition named "[else]" marks the path that should be used if all other conditions evaluate to false. Alternatively, conditions can be placed in square brackets directly at the transition arrow leaving an activity.

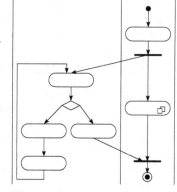

Fig. 2. Activity Diagram for Workflow Models

A path is any valid "way" through a workflow graph along the transition arrows. More than one path can be active at the same time. *Concurrency* is modeled by synchronization bars which are drawn as thick lines. Several transition arrows leaving a synchronization bar model parallel paths which can be executed independent of each other. If several transition arrows point to one synchronization bar these paths are synchronized at this point, i.e., the execution of the following activity is blocked until all paths arrived at the synchronization bar.

The structure of the graph can be mapped to the WfMC control structures [12] as follows: AND-split as well as AND-join are represented by synchronization bars, while an OR-split is modeled by a condition symbol and OR-joins are modeled by several transition arrows pointing to the same activity or condition symbol.

5   Combination of Workflows and Contracts

A service contract should describe the usage and management interactions between customer domain and provider domain. These interactions, e.g., problem management, can be formalized as processes. Processes like production or business processes are modeled with workflow concepts. Therefore, it is reasonable to use these proven concepts to specify service contracts, too. We use workflow concepts in the design phase as well as for the actual contents of the service contract.

Modeled business processes must be understood by people not familiar with workflow concepts. The same is valid for contract design. Users and several experts of customer and provider domain need to be involved in the specification of technical and organizational parts of the contract. Workflow models are easy to understand and therefore a good basis for communication of people with different knowledge. Furthermore, the resulting workflow graphs are a part of the actual contents of service contracts to support the constructive cooperation in the operation phase.

Each process as well as each activity in a process can have quality criteria which must be guaranteed and therefore defined in the contract. But the quality criteria are usually, due to changing needs, much more dynamic than the processes which do not change for a long time. The contract structure needs to reflect the different levels of dynamics because changing quality criteria is common in long–term relationships.

5.1 Contract Structure

This approach divides the contract in three segments which include the elements of the contract identified in section 3.2: legal information, organizational inter-actions and technical functionality as well as service levels. The top level of the contract structure is depicted in figure 3.

Basic agreements represent the legal information in the contract and define this structure. Therefore, the basic agreement is the basis of the contract.

The *service agreement* specifies the processes needed to use and manage the service from the customer's domain. It defines the observable service function-ality and the management facilities, but not including the service level. The processes are service level independent through the use of variable data and vir-tual resources. The service management of the provider is not specified but it must support the agreed interactions.

Service level agreements specify the service level by agreeing upon service capacity and quality as well as upon usage and management interfaces for the customer's domain. This is done by specifying concrete values for the variable data and interfaces for the re-sources of a process. The interfaces must be implemented by the provider, but the way of implementing them is not part of the contract.

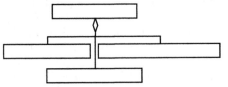

Fig. 3. Structure of a Service Contract

5.2 Process Model

The interactions between customer and provider domain are represented by pro-cesses which are modeled using workflow concepts. Our process model which is compatible to WfMC terms is depicted in figure 4 includes all relevant process elements and shows their relationships.

Processes consist of *activities*. To enable hierarchical decomposition a process itself is an activity. An activity needs *resources* to be executed on and *data* to work with.

Resources are abstract representatives of *objects* and *roles*. Data are either attributes or variables depending on the point in time of the value assignment. *Attributes* are static data which are defined in the contract. *Variables* are highly dynamic values which are exchanged between the domains at run time.

Activities can generate *events*, e.g., for inter–process communication, syn-chronization of concurrency or for exception handling. Events are also *triggers* for processes.

The process execution is controlled by the process *logic*. It includes a static part, the *structure*, as well as a dynamic part, the *conditions*. There are three classes of conditions. *Pre-conditions* and *post-conditions* are used to model constraints for the execution of activities. *Selections* are used by the static structure to select a path by a dynamic decision. Events and data of the process as well as system data like current time can be used in all conditions.

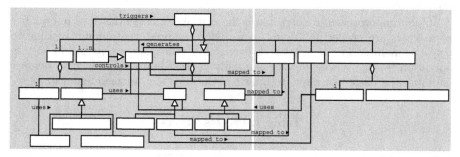

Fig. 4. Process Model

Service agreements contain all this information to model service functionality and management facilities but they do not contain any concrete values for capacity and quality of the service. Service level agreements define this as depicted on the right hand side of figure 4.

Service level agreements map the data needed by activities. Variables are mapped to interfaces to be exchanged at run time, while attributes are mapped to *values* defined in the service level agreement. The resources used by activities are also mapped to interfaces. These *interfaces* hide implementation details. A resource representing a role can be mapped to a technical interface, e.g., defined by a telephone number, or to an individual contact person which is usually done at run time.

An important information in service level agreements are *classification numbers*. In combination with attributes they specify the service level. Classification numbers consist of *expressions* and *assessment schemas* to rate the results. Expressions can refer to variables, attributes and system data as well as events.

5.3   Process Classification

The processes should be classified functionally to deliver an additional level in the structure of service agreements and service level agreements as well as to help identifying all required processes of a service. The process classes shown in table 1 have been identified by combining the TeleManagement Forum's Telecom Operations Map [10] and OSI's Systems Management Functional Areas [3]. The grouping to classes follows a process–oriented contract view mapping processes participating to similar tasks to one class. The class usage completes the classification representing the interactions of the actual purpose of the service.

Some of the processes only exist to allow specification of quality criteria for the service. Classes consisting just of such processes are nevertheless needed in the service contract. The remainder of this section describes the classes in more detail by listing the main processes and typical quality criteria.

Before the service can be used it must be installed. The *provisioning* class includes installation, test, acceptance and if necessary migration processes. Timeliness is an important quality parameter for processes in this class. *Usage* is the most common interaction. Service levels define for example the response time of the service.

Operation processes are invisible to the customer most of the time, but quality parameters are needed for them anyway, for example, the amount of time to detect a malfunction of a network connection is a quality criterion of service operation. As the service quality is usually at least limited during *maintenance* activities the customer wants to limit e.g. the time slot for maintenance.

Table 1. Process Classes

Provisioning	Installation of the service
Usage	Normal usage of the service
Operation	Service management of the provider
Maintenance	Plannable activities needed for service sustainment
Accounting	Collection and processing of accounting data
Security management	Management and observation of service security
Change management	Extension, cancellation and change of service elements
Problem management	Notification, identification and resolution of service malfunction
Contract management	Management of contract contents
Customer care	Service reviews and user support
Deinstallation	Deinstallation of the service

The method used for *accounting* purposes is an important fact because it influences the price of the service. The processes of the *security management* class include mechanisms to observe the service security and to manage the means for authentication of service access.

Minor changes of the service are not uncommon in long–term relationships. This could be the extension of resources or addition of further dealers in the presented scenario. Thus, a *change management* is necessary. A well working *problem management* is a very important factor in outsourcing. Problems can be solved faster if both partners are working together smoothly using well defined processes.

Often, requirements of customers change during contract lifetime. If a change is predictable but not the point in time it is useful to specify *contract management* processes for announcing the need, negotiating and changing the contract. Communication on strengths and weaknesses of the service, its quality or future optimizations should be some of the subjects of regular *customer care* contacts. Besides, a help desk or other support offers are usually necessary for complex services.

Deinstallation processes usually need to be specified if equipment of the provider is installed at customer locations or if support of the provider is needed for migration to the following service.

Additionally, it is useful to define basic processes which are needed in many classes. Such processes are e.g., documentation, feedback and escalation mechanisms.

6   Contract Design

The basis of service contract design are the processes of the customer domain which use and manage the service. The operation processes of the provider must fulfill the contract, but their implementation is not an element of the contract. If a service was implemented by the customer before the current management processes focus too much on implementation issues. Therefore, the management processes of the customer need to be redesigned when outsourcing a service for the first time.

6.1   Design Process

To explain the application of our approach on contract design a simple design process is presented. It shows the necessary steps without detailing their execution. In practice it is an iterative process but the order of the steps in each iteration is the same. The process model explained in section 4 is the basis of this process.

First of all, the process classes needed for the specific service must be identified. Then, the design process depicted in figure 5 is executed for each class. As long as a new process is found (1) the outer iteration (2) is repeated. Hierarchical decomposition of processes is possible because each activity can be modeled as a process.

The left hand side concurrent path (3) defines the triggers of the process (4). The second of the parallel paths (5) builds the process logic with the inner iteration (6). As long as an activity can be identified (7) the activity, the necessary control structures and conditions for the dynamic aspects are inserted in the logic (8). Then, the needed data (9) and resources (10) are specified. Thereafter, the loop is repeated (11) to identify the next activity. When execution of both paths of the outer concurrency is finished (12) the iteration is restarted to identify the next process.

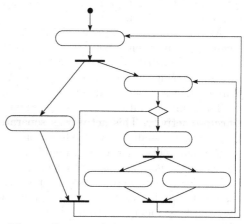

Fig. 5. Design of the Service Agreement

For the specification of the service level agreement the abstract resources, attributes and variables must be mapped. The mapping of resources to specific interfaces can influence the possible service level and depends on the infrastructure of the customer domain. Therefore, the service level is influenced by interfaces, values of attributes and for complexer capacity or quality aspects, which cannot be expressed by a single value, classification numbers must be specified.

Such classification numbers are also needed for quality criteria covering several activities but not the whole process. The expression of the classification number calculates a value for one aspect of the service level. This value is rated by the assessment schema. A schema is just required for guaranteed classification numbers. Some of the classification numbers can be declared dynamic and a process for their change can be agreed upon to enable the controlled adaption of the contract to changing needs or technology during its lifetime.

6.2   Example of a Modeled Process

The design process is demonstrated in a simplified example from the scenario presented in section 2. The example model is a part of the problem management class. It defines the process of fault identification, notification and resolution (see figure 6). The models of the complete process including cooperation during fault identification, documentation, prioritization, exception handling, escalation mechanisms, etc. would go beyond the scope of this paper. During identification of different activities the necessary resources and data must be defined which is also not detailed in this example.

A fault can be detected by both, customer and provider. Therefore, two start symbols exist. Depending on the party that detects the fault the process starts with the *fault detection* on the customer or the provider side. If the customer detects a fault, the next activity is the *notification* of the provider. If the provider detects it the customer is notified by the *customer notification* process. The customer notification is modeled as a composite task because it is a basic process controlled by dynamic values like the message and the receiving role to adapt it to the respective situation.

At this point the two paths join. The next activity is *trouble ticket creation*. Then, the concurrency construct models the fault resolution with regular feedback to the customer on the current state. The feedback path is modeled as a loop monitoring the trouble ticket state. As long as it is open the *customer notification* process is repeated. The feedback interval is realized by a post-condition.

The fault resolution path on the right hand side starts with the *fault identification* activity. This activity is separated from the *fault resolution* because the customer usually wants feedback on the estimated duration and scope of the service malfunction as soon as possible which therefore is modeled explicitly by the *customer notification* base process.

The identification and resolution activities by itself are of no interest to the customer, but he wants to define quality criteria on this activities. Therefore, they must be modeled. After the problem is solved the *trouble ticket* is *closed* which
ends the feedback loop and the customer is notified of the resolution in the final *customer notification* process. The resulting process model including the data and resource definitions is part of the service agreement.

In the next steps the service level agreements are defined by mapping data and resources to values and interfaces. The necessary interfaces are easy to find because each interaction crossing a domain boundary is visible in the workflow graph when the control flow crosses a swimlane boundary. The only new interface added by this process is the fault notification interface which could be mapped to a telephone interface. The interfaces of the composite tasks are defined in the respective workflow model.

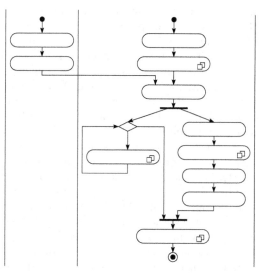

Fig. 6. Fault Notification Process

All requirements for the service level not expressed by now must be specified as classification numbers. In the presented example process a classification number on the total time from trouble ticket creation to trouble ticket closing should be covered by a classification number. Additionally, to limit the amount of service failures, e.g., a classification number for the number of trouble tickets open simultaneously and in certain intervals is needed. Several other classification numbers are thinkable, too.

7   Analyzing the Use of Processes in Contracts

The idea to fulfill the raised requirements is to use concepts of workflow modeling during the design phase and to use the resulting workflow graphs as part of the actual service contract. This is a promising approach because workflow modeling is usually done by gathering knowledge of various sources which is similar to the creation of service contracts. There is not a single expert for all service details of networks, servers, applications, help desks, logistics, etc. The concepts of workflows and the resulting models are easy to understand. Thus, it is possible to model a service contract with workflow concepts while keeping the understandability for people that can give input or have to work with the resulting service contract. The workflow graphs support the information exchange and therefore support cooperation in the design of the service.

The resulting models in the service contract serve as a manual for users and operators. It provides concrete and constructive instructions supporting usage and operation of the service, which improves the cooperation between outsourcing partners. This optimizes the handling of critical situations as it tells affected persons how to become active.

Concepts of workflows have been a research subject for many years now and have already been applied successfully by many enterprises to reengineer and optimize their business processes. So workflow modeling is a proven concept which we are adapting to contract design and writing. Describing the service contract in the same way as the processes in service management also simplifies the mapping of the contract to the provider's service management significantly. Furthermore, the explicit statement of resources, interfaces, competence and information flows supports automation efforts and optimization of resource consumption.

The systematic identification of tasks along the timeline results in more consistent as well as complete and therefore higher quality service contracts. All technical parameters are easier to identify when analyzing the usage process along the timeline.

Furthermore, the process models are highly reusable. They can be reused in higher level processes, further contracts or for other purposes like optimization efforts. Workflow models are precise enough to specify a contract by arbitrarily refining the tasks, but it is also easy to abstract from well known facts and unimportant details. Another feature of workflow models is the possibility to verify them syntactically [8], [4], i.e., to verify the syntactic consistency of a contract specified by a workflow model.

The use of the customer view in form of the customer's business processes results in a customer–centric contract. Additionally, it is service–oriented because the customer's processes focus on usage and not on implementation. The differentiation of the contract in several segments facilitates the controlled dynamic adaption of the contents during contract lifetime. This enables the revision of quality aspects independent from functionality allowing to shorten the negotiation process substantially and therefore enhances the flexibility.

The shortcoming of the approach is that the creation of a contract according to this approach requires lots of effort. But a relevant amount of effort is acceptable for the difficult task of specifying a good contract for a complex service with a contract lifetime of several years.

8   Conclusion and Future Work

This paper proposes the application of workflow concepts for the specification of contracts concerning IT services. The main points of the approach discussed are that it supports the identified requirements for long–term service contracts which need to be customer–centric, dynamic and constructive. Supplementary, this approach generates some additional advantages, like the systematic design of service contracts. The most attractive benefit of the workflow concepts is the active support of cooperation for usage and operation resulting from the instructive nature of workflows. Additional features like controlled adaption are partly enabled by the presented contract structure.

The approach is currently verified in an industry cooperation with a major provider of telecommunication services. The necessary effort sets the main focus especially to extensive, custom–designed services. But we are researching ways to simplify the reuse of this effort.

Further research is concerned with substantiating the application methodology and extraction of characteristic capacity and quality parameters. The target is to identify a methodology for the systematic derivation of customer–oriented but measurable quality parameters.

Acknowledgment

The author wishes to thank the members of the Munich Network Management (MNM) Team for helpful discussions and valuable comments on previous versions of the paper. The MNM Team directed by Prof. Dr. Heinz-Gerd Hegering is a group of researchers of the University of Munich, the Munich University of Technology, and the Leibniz Supercomputing Center of the Bavarian Academy of Sciences. Its web–server is located at
`http://wwwmnmteam.informatik.uni-muenchen.de`.

References

1. P. Bhoj, S. Singhal, and S. Chutani. SLA Management in Federated Environments. In M. Sloman, S. Mazumdar, and E. Lupo, editors, *Integrated Network Management VI (IM'99)*, Boston, MA, May 1999. IEEE Publishing.
2. H.-G. Hegering, S. Abeck, and B. Neumair. *Integrated Management of Networked Systems – Concepts, Architectures and their Operational Application.* Morgan Kaufmann Publishers, ISBN 1-55860-571-1, 1999. 651 p.
3. Information Technology – Open Systems Interconnection – Systems Management Overview. IS 10040, International Organization for Standardization and International Electrotechnical Committee, 1992.
4. A. Karamanolis, D. Giannakopoukou, J. Magee, and S. Weater. Modelling and Analysis of Workflow Processes. Technical Report DTR99-2, Imperial College of Science, 1999.
5. Business Process Modeling with UML. TC Document ad/00-02-04, Object Management Group, February 2000.
6. Unified Modeling Language (UML) 1.3 specification. OMG Specification formal/00-03-01, Object Management Group, March 2000.
7. T. Preuß and H. König. Service Supplier Relations for the Outsourcing of Information Processing Services. In *Proceedings of the IEEE Enterprise Networking and Computing Conference (ENCOM 98)*, Atlanta, GA, USA, 1998.
8. W. Sadiq and M. Orlowska. Modeling and Verification of Workflow Graphs. Technical Report 386, University of Queensland, November 1996.
9. W. Sadiq and M. Orlowska. On Capturing Process Requirements of Workflow Based Business Information Systems. In *Proceedings of the 3rd International Conference on Business Information Systems (BIS99)*, Poznan, Poland, April 1999.
10. SMART TMN Telecom Operations Map. Evaluation Version 1.1 GB910, TeleManagement Forum, April 1999.
11. The Workflow Reference Model. TC Document 00-1003, Workflow Management Coalition, January 1995.
12. Terminology & Glossary. TC Document 1011, Workflow Management Coalition, February 1999.

Using Time over Threshold to Reduce Noise in Performance and Fault Management Systems

Mark Sylor[1] and Lingmin Meng[2]

[1]Concord Communications, 600 Nickerson Rd. Marlboro, MA 01752
sylor@concord.com

[2]Department of Electrical and Computer Engineering, 8 St. Mary Street
Boston University, Boston, MA 02215

Abstract: Fault management systems detect performance problems and intermittent failures by periodically examining a metric (such as the utilization of a link), and raising an alarm if the value is above a threshold. Such systems can generate numerous alarms. Various schemes have been proposed for reducing the number of alarms, or filtering out the important ones. The time over threshold detection algorithm reduces the volume of alarms at the source detector. This paper describes an experiment that compares time over threshold against simple threshold crossings. The experiment demonstrates that it reduces the number of alarms raised by a factor of 25 to 1 without any significant reduction in the problems detected.

1 Introduction

Concord s Network Health" product, a member of the eHealth" product suite, collects performance and fault management information from networks, systems, and applications. It stores the information for historical analysis, and presents the information in report format on the web or on paper. The information collected is used to analyze the overall health of the networks, systems, and applications, and to support capacity planning.

Concord has recently added LiveHealth to the eHealth product family. LiveHealth analyzes the information Network Health already collects in real time to detect faults and performance problems. Because Network Health maintains a historical record of past performance and faults, it was natural to use that data to improve the detection capabilities of LiveHealth.

Most fault and performance management systems depend on simple threshold crossing events to detect problems. They periodically sample the value of some performance or fault metric, and compare it with a fixed value, called a threshold. If the value of the metric is greater (or less) than the threshold, an alarm is raised. When the alarm is raised, a notification is sent to a network management system (an NMS), generally in the form of an SNMP trap, a CMIP notification, or in some proprietary format.

While simple threshold crossings are effective in detecting problems, they generate far too many alarms. The metrics indicative of performance problems show a wide variation, with little predictability in their pattern. For example, Fig. 3, shows a graph of

A. Ambler, S.B.Calo, and G. Kar (Eds.): DSOM 2000, LNCS 1960, pp. 145 - 156, 2000.
© Springer-Verlag Berlin Heidelberg 2000

the utilization of a link over three days measured at 5 minute intervals. As is well known, such metrics vary widely. Other metrics, discarded packets, error rates, congestion indications, disk I/O rates, CPU Utilization, application workloads (transaction rates), all show similar high variation. This variation guarantees that some samples will be above any achievable threshold. No matter how high the threshold is set, sooner or later, that threshold will be exceeded and an alarm will be raised. These alarms are a form of false alarm.

False alarms have a serious impact on any real time fault or performance management system. If there are too many alarms, operators will tend to ignore them all, including the alarms that indicate real problems. Even if the operator is conscientious, finding a particular alarm from a list of thousands of alarms is difficult.

One approach to dealing with this flood of alarms is to filter, classify, and prioritize the alarms in the network management system receiving the notifications. For example, an NMS might filter out unimportant alarms based on a severity field included in the notification. An NMS might classify the alarms based on fields in the notification, the element (object, or host) raising the alarm, the type of alarm, the variable which exceeded the threshold, and other fields. Once classified, an NMS might simply count the number of events of a class. Based on the count or the class, the NMS might prioritize the alarm, or take action on the alarm. The kinds of actions an NMS might take include actions to notify an operator (for example by paging the operator). Another is to change the state of the NMS, for example, receiving an alarm of a particular class causes a change in the rules so subsequent alarms of that class are discarded. While all these are appropriate actions, they can be difficult to set up, and may not reduce the overall alarm rate.

With LiveHealth, we attempt to decrease the flood of alarms at the point of detection, rather than provide a better means of handling a flood of alarms generated by the detector. The technique adopted for reducing alarms is based on a heuristic detection algorithm called *time over threshold*.

2 Related Work

A common technique used to decrease the flood of alarms has been to use thresholds to drive an alarm state. When the threshold is exceeded, an alarm is raised. When the value falls below a threshold, the alarm is cleared. In some of these systems, the falling threshold can have a different value than the rising threshold. This technique was standardized in [5], implemented in commercial Network Management Systems such as [3], and implemented in agents within network devices such as [6]. While these techniques help, experience shows they do not reduce the alarms enough. Further, they depend on setting the falling threshold correctly, yet there is no obvious value that is appropriate.

More recent work has focused on using statistical approaches to setting the threshold [7], or in using statistical techniques to detect points of change that may indicate a fault [2]. These approaches improve the quality of alarms by more accurately predicting when an alarm is likely. Part of the implementation of Live Health exploits very similar techniques to set better thresholds. This work is not covered in this paper. Our experience so far is that these approaches do not lower the alarm rate, in fact they tend to increase the number of alarms, as they detect problems that previously were missed. Approaches combining improvements in detecting problems (such as statistical thresholds) with good noise reduction techniques such as described here are needed to provide users with high quality alarms.

3 Operation of Network Health

Network Health periodically polls counters from elements (managed objects) in the network. It computes and stores the differences in the counters between the samples, and the difference in time between the samples, which is stored in the database as a *statistics record*. From the statistics record, performance and fault metrics (called trend variables in Network Health) can be computed. These trend variables are often rates, such as *bytes sent per second*. The value of a rate trend variable is the average rate over the interval covered by the sample period. From these basic statistics records, and trend variables, Network Health computes numerous reports on the performance of the network, systems, and applications.

With the addition of Live Health, Network Health takes those same statistics records, and passes them to an evaluation engine called the LiveExceptions Engine (LE engine) in addition to, or instead of, the database. The purpose of the LE engine is to detect performance or fault problems. The LE engine detects problems by evaluating current metrics against a set of rules. Live Health includes a rich collection of default rules. When a rule detects a problem, the LE engine raises an alarm. When the rule detects that the problem has gone away, the alarm is cleared. Alarms are displayed in the Live Exceptions alarm browser. An alarm can be sent as a trap to an NMS where it is displayed with other events, or can be used to drive the status (color) of objects in the NMS map.

4 Time over Threshold

The time over threshold algorithm is implemented in the LiveExceptions Engine. When a statistics sample is received for an element the engine analyses the statistics against all the applicable rules. Each rule defines a detection algorithm to apply to the data, and any parameters used to control the algorithm. The time over threshold algorithm computes the value of a trend variable over a time period called the analysis window, and compares it with a threshold. It then determines how long the variable was over the threshold. If that the variable was over threshold for a length is greater than an alarm window, then an alarm is active for the sample period. When an alarm changes from inactive to active, we say the alarm is raised. When an alarm changes from active to inactive, we say the alarm is cleared.

A typical example is an alarm on the CPU Utilization of a Unix server. We raise an alarm if the CPU Utilization is greater than 90% (the threshold) for more than 15 minutes (the alarm window) out of the past hour (the analysis window).

More formally, the time over threshold algorithm can be defined as follows.

Let R be a time over threshold alarm rule defined for a trend variable $x(t)$, where the rule defines a threshold, T, an analysis window, W, and an alarm window A.

Assume at time t_n the engine receives a new statistics sample x_n of the trend variable $x(t)$ and this sample covers the period $t_{n-1} < t \le t_n$, i.e., for all $i = 1..n$

$$x(t) = x_i \text{ for all } t \text{ such that } t_{i-1} < t \le t_i$$

Further, assume that the samples x_i for $i = j..n$ cover the analysis window, i.e.,

$$t_{j-1} < (t_n - W) \le t_j < t_{j+1} < ... < t_n$$

Let the threshold state of sample i, c_i represent whether x_i exceeds the threshold, c_i be defined as

$$c_i = \begin{cases} 1, & x_i > T \\ 0, & \text{otherwise} \end{cases}$$

Then we compute, b_n, the time $x(t)$ is over the threshold T in the analysis window W for sample n as

$$b_n = (t_j - (t_n - W))c_j + \sum_{i=j+1}^{n} (t_i - t_{i-1})c_i$$

Finally, let a_n represent the alarm state of a rule for sample n be defined as

$$a_n = \begin{cases} 1, & b_n \ge A \\ 0, & \text{otherwise} \end{cases}$$

We use the alarm state to raise and clear alarms as follows. If $a_{n-1} = 0$ and $a_n = 1$ then we raise an alarm at time t_n. If $a_{m-1} = 0$ and $a_m = 1$ then we clear that alarm at time t_m. If $a_n = 0$ we say the alarm is inactive, if $a_n = 1$, we say the alarm is active.

This definition describes the basic idea behind time over threshold. We have generalized the definition in two ways.

First, the time over threshold computation of the condition state can use more general expressions to compute the threshold state. Live Exceptions supports expressions such as

$$(x_i < T)$$
$$((x_i > T) \& (y_i > S))$$

Live exceptions also supports dynamically computed thresholds based on long term historical analysis of the behavior of the variable. That part of the work is not covered here.

Second, the samples may not perfectly cover the analysis window. Failures in the polling process or in the devices being monitored can lead to gaps in the data. When the system initially starts monitoring a rule, there is some start up period where the samples will only cover a portion of the analysis window. In these cases, the threshold state $a(t)$ of any period that is uncovered by a sample is assumed to be 0 (false).

The time over threshold algorithm is similar to, but not the same as an algorithm based on the number of samples over the threshold. We used time rather than samples for a number of reasons.

- Samples cannot be collected on precisely regular intervals. A small jitter in the time between polls on the order of a few seconds is introduced because:
  - The network introduces delays in packet latency.
  - Other activities on the system running Network Health add jitter to the sampling process.
  - Other activities in the system being monitored generate jitter in the sampling periods.
- Polls may be lost.
  - Communications problems in the network can cause SNMP requests or responses to be lost.
  - Agents may sometimes fail to respond. An agent may have limited memory to buffer requests or have other limitations that cause them to drop SNMP requests.

  Network Health can and does recover the data for those missed polls. When it does recover the data, the resulting sample covers two or more of the scheduled sample periods.

For these reasons, using the time each sample covered, rather than the number of samples collected is a better measure of behavior.

The time over threshold algorithm is designed to handle a number of common behaviors in trend variables.

Many variables experience isolated spikes such as those in Fig. 4 that graphs the percentage of frames which had an error on a WAN link. In many cases, a single, isolated spike is not a real problem. Only when enough samples are bad should an alarm be raised. By setting the alarm window, A, to a longer period (say 15 minutes), we can ensure that an alarm is raised only when the problem persists long enough to impact the system.

When a variable crosses over a threshold, there are likely to be periods when the variable will fall below the threshold for a few samples, only to return above the threshold shortly thereafter. The analysis window (or more precisely, the analysis window, less the alarm window) controls how long the variable must remain below the threshold before the alarm is cleared. In general, increasing the analysis window reduces the probability that when an alarm is cleared, it will simply be raised again within a short time. Fig. 5 shows a typical situation where a variable crosses above a threshold and stays above for most of the time, but occasionally falls below the threshold. In this case the time over threshold raises the alarm at the beginning of the problem, and keeps it active throughout the period.

5 Experiment

As we were developing the set of default rules used to detect problems we ran numerous experiments on many live networks, and on saved databases of data collected by Network Health. We developed a tool to replay a database by reading the collected samples, and passing them to the LE engine as if they had been polled. This tool allowed us to compare the behavior of different rule sets, and to fine-tune the thresholds and parameters that control the rules. The tool also allowed us to evaluate and compare the detection algorithms with more conventional methods against data collected from real networks. One of these experiments is reported here.

The database used in the experiment covers a period slightly more than three days of monitored data. Each element was polled at a 5 minutes rate. The number of samples for each rule is approximately proportional to the time duration, in this case about 900 samples. 1274 elements were monitored, including networks, systems, and applications. The types of elements monitored included: 43 routers or switches, 9 servers, each running from 1 to 3 applications, 5 Network Access Servers, 69 frame relay circuits, 68 LANs, 326 WAN links, and 2 ATM channels. Each element was evaluated against the complete set of default performance and fault rules. Each element had from 5 to 15 alarm rules applied to it. The network is a fairly typical enterprise network. Although only a portion was monitored, the elements were representative of the whole network.

The goal of this experiment was to compare the detection effectiveness of three algorithms: Time over Threshold (TOT), Threshold Driven State (TDS), and Simple Thresholds (ST).

The Simple Thresholds (ST) algorithm simply tests the sampled trend variable against the threshold, and raises an alarm whenever the variable is above the threshold. Because a trap is sent for every sample over threshold, it does not send alarm clear traps. One apparent problem with ST is that it raises so many alarms that the important information to the system manager could be buried in the flood of alarms. The simple threshold forms a basis of comparison that any algorithm can be evaluated against.

The Threshold Driven State (TDS) algorithm attempts to compensate for the problems in ST by remembering the state of the alarm for the previous sample, and sending a trap when the state changes. The TDS algorithm can reduce the number of alarms raised for when an alarm is caused by a consecutive sequence of bad polls, such as seen in Fig. 5. However, it has problem when the value bounces up and down crossing the value threshold frequently such as that shown in Fig. 6. Variables with high variability around the threshold cause TDS alarms that babble. One fix to TDS that has been proposed is to use a lower threshold to clear the alarm than the threshold used to raise the alarm. It is difficult to find a good clearing threshold. Consider for example the system shown in Fig. 7, here TDS would raise an alarm on each of the spikes over the threshold, and clear it on the next sample. Each alarm would be a false alarm, and no reasonable setting of a falling threshold would correct that problem.

To compare the three algorithms, we ran two replays of the database. First we ran the default rules using the standard TOT windows against the database. Most of these rules use an alarm window, A, of 15 minutes, and an analysis window, W, of 60 minutes.

For the second run, observe that if the TOT algorithm is run on rules where $A = W$ is less than the minimum sample period, then by the definitions above, the alarm state a_n equals the threshold state c_n of the sample. Since both the ST and TDS algorithms are based on the threshold state, we can reconstruct how the algorithms will behave. In particular, by setting both the analysis window size and the alarm window size to 1, we are able to reconstruct the original binary information whether the variable at each time interval is above or below the threshold. We then processed this binary information to determine the behavior of the ST and TDS algorithms.

6 Results

We compute the following 5 performance parameters for each algorithm to determine their effectiveness at detecting problems.

1. The number of bad samples (samples when the monitored variable is above the threshold) covered by the raised alarms.

2. The number of good samples (samples when the monitored variable is within the normal range) covered by the raised alarms.

3. The number of alarm set traps each algorithm sends.

4. The number of alarm clear traps each algorithm sends.

5. The total number of traps (both alarm set and alarm clear) each algorithm sends. The traps draw the network manager or administrators attention. This number should be as small as possible to reflect only the real network outages or potential problems.

The following table summarizes the performances of the three algorithms, for all the elements and all of the rules.

Table 1. Comparison of Algorithms

	Simple Thresholds (ST)	Threshold Driven State (TDS)	Time Over Threshold (TOT)
# bad polls covered	17505	17505	16373
# good polls	0	0	10131
# alarm set	17505	3836	709
# alarm clear	0	3836	709
# traps (set or clear)	17505	7672	1418

The ST and TDS algorithms raised alarms for 223 elements, while the TOT algorithm raised alarms for 158 elements. An example of a case where TOT did **not** fire an alarm is that shown in Fig. 4. These isolated spikes were a source of many false alarms.

The TOT algorithm covered 93.5% (16373/17505) of all the bad polls, or equivalently, 93.5% of the time when the variable is above the value threshold. The number of alarm sets was reduced by a factor of 25 (17505/709) from ST or 5.4 (3836/709) from TDS.

Note that alarms raised by the TOT algorithm covered a small number of good polls (10131). These polls lie in the gaps between bad polls and serve to reduce the total number of raised alarms. This number also indicates the frequency of bad polls during the alarm periods raised by the TOT algorithm. The frequency of bad polls vs. good polls during these periods is 8:5 (16373:10131).

The number of traps is reduced by a factor of 12.3 (17505/1418) from ST and 5.4 (7672/1418) from TDS.

Fig. 1. Comparison of Algorithms

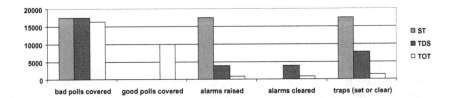

We also examined the behavior of other related algorithms. Roughly, the TOT algorithm can be viewed as a rectangular window filtering of the binary sequence of a monitored variable (1 if sample value over threshold, 0 if below threshold). In an effort to smooth the result, we also tried two other types of window filtering. One is the exponential forgetting filtering, which uses an infinite length window of all available data by putting exponentially attenuated weights on historical samples. The other is a Gaussian shaped window, which has same length as in the TOT but puts more weight on most recent sample and less weight on the past samples. Similar approaches have been used in [1].

The results of these three window filtering are compared in Figure 2. It shows the behavior of a single rule for a single element. The X-axis shows the time, and the Y-axis shows the number of bad polls in an analysis window of width 12. As is shown, these two window alternatives did not improve the TOT algorithm in terms of smoothness and latency. In fact, these two window alternatives gave less smooth results due to non-integer operations. Because these alternative filters are more difficult to explain, and gave no better result, we chose to stick with the simple TOT rectangular filter.

Fig. 2. Comparison of Alternative Window Filtering

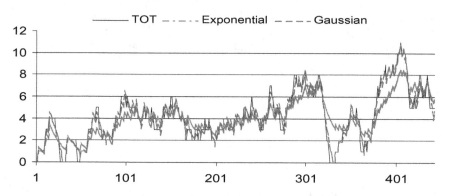

7 Conclusion and Future Work

The time over threshold detection algorithm does a good job of reducing the number of alarms raised, and therefore the number of traps that must be processed by an NMS. Yet it does not reduce the ability to detect problems. By far the majority of isolated spikes are transient conditions, which are not indicative of problems. We have implemented algorithms that dynamically determine thresholds based on a statistical analysis of historical behavior. While that work is not covered in this paper, we believe it to be a fruitful area for further research. We certainly have not examined all of the noise reduction algorithms that might be applied to performance and fault management. We believe that any algorithms proposed must be evaluated against real world data such as used in this work.

Of course the true test of any problem detection system is field experience in detecting real problems in real networks, systems, and applications. Our experience with Live Health to date in these real world networks indicates that the basic Time Over Threshold algorithm is effective at reducing noise.

8 Figures

Fig. 3. Typical Variation, Outbound Utilization on a 128 Kbit/sec Link

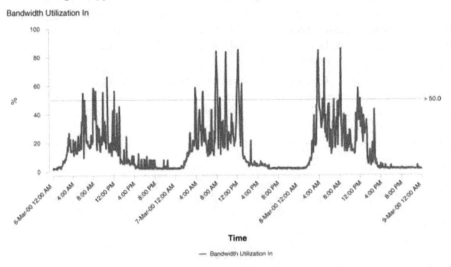

Fig. 4. Errors on a 64Kbit/sec WAN Link

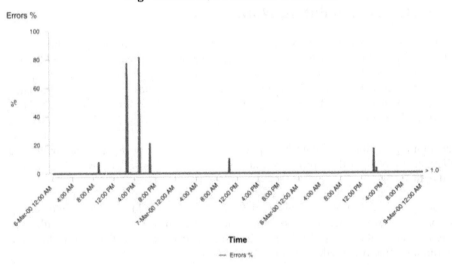

Fig. 5. Outbound Utilization of a 128 Kbit/sec WAN Link

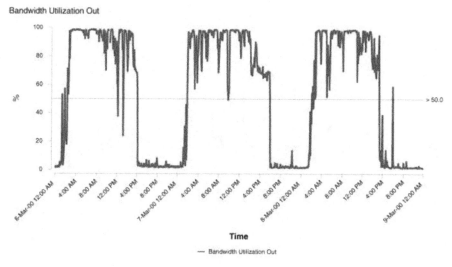

Fig. 6. CPU Utilization too High Alarm that Causes TDS Babbling

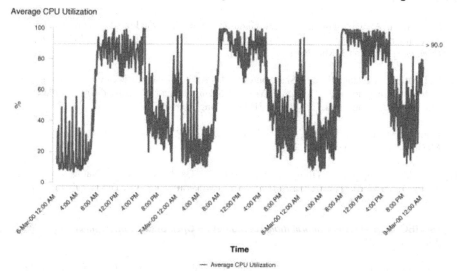

Fig. 7. CPU Utilization of a Unix Server

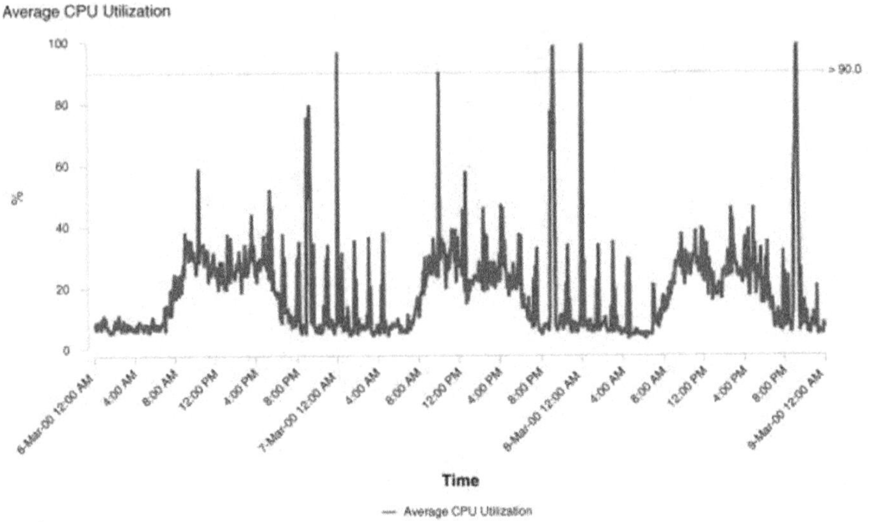

Average CPU Utilization

References

[1] Bondavalli, A., Chiaradonna, S., Di Giandomenico, F., Grandoni, F., Threshold-Based Mechanisms to Discriminate Transient from Intermittent Faults , *IEEE Trans. on Computers*, v 49, no 3, Mar 2000, pp. 230-245.

[2] Hellerstein, J. Zhang, F., and Shahabuddin, P., An Approach to Predictive Detection for Service Management , *Integrated Network Management VI,* Edited by Sloman, M., Mazumdar, S., and Lupu, E., 1999, IEEE Publishing, pp. 309-322.

[3] Huntington-Lee, J., Terplan, K., and Gibson, J., *HP OpenView: A Managers Guide,* McGraw-Hill, New York, NY, 1997, pp. 137-9.

[4] Lelend, W., Taqqu, M., Willinger, W., Wilson, D., On the Self-Similar Nature of Ethernet Traffic (ExtendedVersion) , *IEEE/ACM Trans. on Networking*, v. 2, no.1, Feb 1994, pp.1-15.

[5] ISO/IEC 10164-11:1994 *Information Technology — Open Systems Interconnection — Systems Management: Metric Objects and Attributes.*

[6] Maggiora, P., Elliott, C., Pavone, R., Phelps, K., and Thompson, J., *Performance and Fault Management*, Cisco Press, Indianapolis, IN, 2000, pp. 91-97.

[7] Thottan, M., and Ji, C., Adaptive Thresholding for Proactive Network Problem Detection , *Third IEEE International Workshop on Systems Management*, Newport, RI, Apr 22-24, 1998, pp. 108-116.

The Doctor Is In: Helping End Users Understand the Health of Distributed Systems

Paul Dourish[1], Daniel C. Swinehart[2], and Marvin Theimer[3]

[1] Dept Information and Computer Science, University of California Irvine, Irvine, CA 92697
jpd@ics.uci.edu
[2] Xerox Palo Alto Research Center
3333 Coyote Hill Road, Palo Alto, CA 94304
swinehart@parc.xerox.com
[3] Microsoft Research, One Microsoft Way, Redmond WA 98052
theimer@microsoft.com

Abstract. Users need know nothing of the internals of distributed applications that are performing well. However, when performance flags or fails, a depiction of system behavior from the user's point of view can be invaluable. We describe a robust architecture and a suite of presentation and visualization tools that give users a clearer view of what is going on.
Keywords: Distributed systems, visualization, diagnostics, management.

1 Introduction

Ideally, users should not have to concern themselves with the internal workings of an application. But most modern personal computer applications are composed from distributed networks of computation, many systems working together to cause the effects the user sees. As complexity and geographic scope increase, it becomes increasingly difficult to answer the questions: "What's happening here?" and "What should or can I do in response?" Current systems typically try to mask faults and performance slowdowns, but leave the user hanging helplessly when they do not fully succeed.

System administrators possess all manner of tools to reveal the behavior of hardware and software systems. The information these tools present is generally focused on the individual components of the system (disks, servers, routers, processes, etc.), and on measures that evaluate the overall health of the environment, rather than end-to-end user applications, distributed services, and the like. Moreover, the information provided by these tools is typically detailed technical information, such as dropped packet counts, throughput characteristics, CPU load and so forth. We believe that the behavioral information that these tools measure could be of considerable value to end users as well. If presented in a focused context and in a way users can comprehend, such information can go far in distributed settings to explain the relationship between system action and the user's immediate tasks.

A. Ambler, S.B. Calo, and G. Kar (Eds.): DSOM 2000, LNCS 1960, pp. 157–168, 2000.

Consider a commonplace example: ordering books online from home. At some point, the process abruptly stops cold. What caused the failure? The local machine, the modem, the phone line, the ISP or backbone router, or the vendor's server? How can an end user understand enough of what is going on to make an informed decision about how to proceed or which customer support line to call? (Depending on the nature of a problem, system administrators may be either unaware of it or unconcerned about the severity of an isolated anomaly, unless alerted by a well-informed user.)

Although particularly noticeable, failures are not necessarily the most valuable behaviors to unravel for users. In particular, properly-functioning but overloaded systems can mimic actual failures, but the most appropriate user response (wait out the rendering of a web page, select a less loaded printer or file mirror site, or simply come back later) is entirely different. If accurate completion estimates can be obtained, a user can make even better decisions. In other words, it is lamentable but true that users continually need to understand system behavior and adjust their own activity to match it. How can we develop systems and tools that can supply these kinds of insights to end users?

This is the question we have addressed in our "Systems Health" project. Its goal is to create an infrastructure for collecting, processing, distributing and presenting to end users information about the ongoing "health" of individual distributed applications: their current status, their overall performance, historical patterns of activity, and their likely future state.

This goal presents a number of challenges. Traditional monitoring tools show "vertical" slices through the system, showing one component in considerable detail and "drilling down" to see more specifics. They collect information that both notifies administrators of an individual problem and provides them with enough information to solve it.

Our concerns are somewhat different. End users need specific, relevant, contextualized information about the components they are actually using and how these components combine to accomplish their tasks. As we have discussed, there may not actually be a "problem" to be solved; the goal may instead be simply to assess the system's state, providing the user with options. For example, a user is likely to be more interested in locating an alternate functioning printer than in knowing how to restore a failed printer to operation.

Making sense of the current state of a system often requires access to historical information. Often, the only way to adequately assess some information is in terms of its variations over time.

In a distributed application, the needed information does not reside in one place. To characterize the behavior of a user's "job" may require gathering remote server performance information, end-to-end network congestion information, and data about local workstation activities, then presenting the information in a form that does not directly match the system's actual structure.

The last thing a user needs is diagnostic tools whose robustness is inferior to the systems they are evaluating. Since many health components will of necessity operate on the same equipment as the observed systems, It is critical that these components operate with minimal impact on the underlying applications, that they fail significantly less often, and that they continue to operate even when major parts of the system have failed.

In summary, our system must possess a number of characteristics:

1. *User-centered.* Our primary concern is to give information relevant to the needs of a system's end users and expressed in terms they understand.
2. *End-to-end.* Most real work tasks involve a variety of components throughout the system. We must describe all those components acting in concert.
3. *Adjustable.* Users need ways to take different routes through this information, and the system must respond to changes in user concerns. Gathering the needed data should be an automatic consequence of user interest.
4. *Historical.* Since some situations can only be understood in context, we need to be able to provide information about the past as well as the present.
5. *Robust.* The system is intended to help users understand problems that may disturb their work. It is critical that those problems do not also interfere with the functioning of the health system. We refer to this criterion as "fail last/fail least"; it drives a considerable amount of our design.

2   Related Work

Software tools to monitor system and application behavior developed in a rather ad hoc fashion, as commands to extract information collected by the modules that make up a software system. The introduction of the "/proc" virtual filesystem in Eighth Edition UNIX [8] was an early attempt to systematize access to active system information.

SNMP, the Simple Network Management Protocol, introduced a coherent mechanism for structuring and accessing management information across a network [13]. SNMP tools remotely query and control hierarchical stores of management information (Management Information Bases, or MIBs). SNMP underlies many integrated tools for distributed system monitoring and management, such as Unicenter [2]; recent enhancements extend to the inner workings of host systems, services and processes. Similar mechanisms, now operating under the umbrella of the Desktop Management Task Force [3] include Web-Based Enterprise Management, the Desktop Management Interface, and Windows Management Instrumentation. Simpler, if less ambitious, freeware or shareware systems include Big Brother [9] and Pulsar [5], designed to monitor an enterprise and inform administrators of problematic conditions. NetMedic [7] does cater to the end user, using passive analysis of Internet traffic to provide explanations for some performance mysteries of Web browsing.

Planet MBone [10] is a research effort to provide visualizations of diagnostic information, simplifying understanding of the Internet's multicast infrastructure. Similarly, software visualization (e.g. [14]) and algorithm animation [1] prototypes apply graphical techniques to provide diagnostic views of the structure and behavior of software systems.

Dourish [4] has argued the value of visualizations as an aspect of normal interaction with software systems, to provide "accounts" of their operation similar to the physical embodiments of everyday objects. Here, a view into a program's behavior is not provided for debugging, but rather to support the ongoing management of user activity. One implication of this perspective is that a system's account of its current behavior should meaningfully reflect the user's tasks rather the system's structure, echoing our end-to-end "horizontal slice" approach.

3   Architecture

Our health system must satisfy both the interactional needs of the system and the need to provide explanations to end-users. The architecture we have designed to meet these requirements is three-tiered. At the top, a variety of *health applications* both collect and display information about the system's health. In the middle is the *Health Blackboard System*, the abstract model to which applications are written; the blackboard is in turn supported by the *Health Repository*, a storage layer tailored to our specific needs.

3.1   Health Blackboard System

Our architecture begins with a relatively conventional agent/database approach. A common information store, the *blackboard*, acts as an indirect, persistent communication channel among any number of individual processing elements, or agents. Agents produce or consume information by writing to or reading from the blackboard. Our architecture (Fig. 1) is built around three different sorts of agents: *gatherers*, *hunters* and *transducers*.

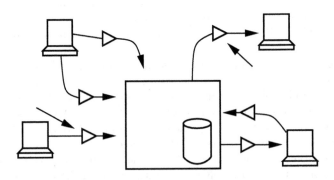

Fig. 1. Hunter and Gatherer Processes Add, Remove Health Data from a Blackboard

Gatherer agents are responsible for gathering different pieces of information from the distributed environment and keeping it up-to-date, by adding or replacing items on the blackboard. Running on or near the computers where the information is generated, gatherers enter onto the blackboard information from a variety of sources: SNMP interfaces, application management interfaces, or system-specific performance queries, allowing the remainder of the system to be independent of the precise collection mechanisms. Proxy gatherers can be deployed to gather data, using other protocols, from hosts that are not able to fully participate in our architecture, and record the results in the blackboard.

Hunters are the components within presentation applications by which information flows out of the system again. Hunter agents consult the blackboard for health information to process and present in some sort of visualization to a user. They function either by searching explicitly for the desired information or by scheduling notification events for new information matching their interests.

Transducers act both as hunters and as gatherers. A transducer is created to process or summarize raw data into items of greater value to the user. It extracts relevant items from the blackboard, adding new items representing the higher-level account of the system's behavior.

Permanently deployed gatherers monitor specific system components on each participating host, in order to maintain a historical record of base level performance information. Most, however, are created when the information they gather is needed. When a hunter asks for information about, say, a particular server load, a *Gatherer Factory* process on the server host creates a gatherer for that information; the hunter either waits for the first results or schedules future notifications. The blackboard will request the removal of a non-persistent gatherer when there is no longer any hunter interested in that information.

Thus, health data enters the system via gatherers, flows via the blackboard through some number of transducers before playing a part in the reports of one or more hunters. (Data representing the load on a network link might be used to generate reports for the many applications that make use of that particular link.) Multiple timestamped items may be inserted concerning the same measured quantity in order to produce a historical record, where needed. Throughout the process, the blackboard controls the dynamics of the system, presents a uniform model for the collection and management of health information, and connects information providers with information seekers.

Our blackboard is implemented as a tuple-space, not unlike that provided by the widely available JavaSpaces [6]. For consistent interoperation between hunters and gatherers, we employ a conventional structure for tuples. Each data object is accompanied by the host, system, subsystem and component that generated it, a timestamp, a name and optional parameters. The data can be any serializable Java object, permitting the system to store and manage information from a wide range of sources, information whose form may change over time.

As a simple example, consider a hunter that maintains a server's load average graph. The hunter initially indicates its interest in this information by putting onto the blackboard a request tuple of the form `<server1, system, cpu, *, load, *>`, thus registering its interest. Any matching tuple is delivered immediately to the hunter; otherwise, the blackboard directs the *GathererFactory* on *server1* to create a new instance of the *LoadGatherer* class which is registered as a supplier of `cpu load` data. The new gatherer periodically adds server load records to the blackboard, generating events that will waken any waiting hunters, including the original one, which now can extract the load information and present it to the user. If the LoadGatherer instance cannot directly generate the information, it may serve as a transducer, recursively extracting additional blackboard tuples to produce its result. Should this hunter go away or become unreachable, the blackboard will delete the pattern record. Periodically, another process garbage-collects gatherers whose information is no longer needed.

This example highlights several aspects of our approach. The blackboard separates the collection of information (gatherers) from the routing of information to interested processes (hunters). This indirection allows us to interpose other agents that process low-level records into higher-level information by combining information on the blackboard. It also allows us to limit the network traffic and processor load involved in collecting and collating information. Finally, health system components persist in the system only while they are needed.

We have stated that historical information is often of considerable importance in understanding the state of a system. Retaining data records for extended periods can provide the same vital function that conventional system logs achieve, with the added advantage that the resulting tuples can be organized through various structured queries. In our current prototype, historical records can be retained by adding multiple timestamped tuples matching the same search specification to the blackboard. Further, through an extensible "hints" mechanism, clients can specify degrees of liveness of the information they require, "time to live" hints for information that is added, and so forth. While this does not yet address all the historical requirements, it provides enough flexibility to allow basic management of the temporality of near-live data.

3.2   Health Repository

Although the Blackboard abstraction can be supported by very simple, centralized, in-memory information stores, our robustness requirements create additional criteria for a storage layer. First, health information must be available with low overhead. Since not only error states but also ongoing performance information must be presented to the user on a continuing basis, we must avoid server bottlenecks and the overhead of synchronous network requests to fetch and store data. Second, the health system must not itself appear to fail as a consequence of other problems. For example, in the case of a network partition, the system must continue to function as best it can. Third, logging requirements argue for storage of diagnostic information beyond the capacities of individual workstations or application servers. Finally, health reports from a large enterprise could overwhelm a centralized server.

These criteria suggest the use of a replicated data store. By storing health information in persistent, replicated repositories, gatherers may collect and report their findings to local replicas, confident that the data will eventually reach those that need it, even in the face of network congestion or partitioning. Similarly, hunters can report the latest information they were able to receive; they can use discontinuities in timestamps to identify the approximate point of failure along a broken network path. We can present the appearance of a system with very low probability of total failure.

But a fully replicated store will also not scale. This argues for partial replication, where information is replicated only where needed, according to the patterns of information requests. Ideally, local replicas should be readily available both to the gatherers that create the information and to the hunters that use it. Furthermore, for information that is of longer term value, long-lived replicas should be retained by servers charged with maintaining historical data. Additional replicas can be designated for further protection against information loss. This approach implies that some amount of data inconsistency can be tolerated, if necessary, in order to keep functioning. For system health information, we need data to be accurate when possible, but slightly out of date information is better than none and can be tolerated when the latest values are unavailable.

Based on these considerations, we chose to adapt the Bayou system [17] as our repository layer. Bayou is a weakly-consistent replicated database, developed originally to support ubiquitous computing environments, where mobile devices routinely experience unpredictable connectivity. Bayou provides mechanisms for dealing with network disconnection and carrying on processing, including *session guarantees* (customizable degrees of consistency management) and *mergeprocs* (application specific code to resolve conflicts after reconnection) [16].

Bayou's replication supports our requirements for high availability. Its weak consistency adequately supports continued operation in the face of temporary failures and network partitions. However, its fully general algorithm for resolving update conflicts is more heavyweight than we need. Health information has specific dynamic patterns of information generation and modification, which we can exploit in designing a replication update scheme. We developed a variant of the original implementation, enhanced for rapid (sub-second) dissemination of updates to all replicas, to underlie our blackboard abstraction.

3.3   Infrastructure Implementation

We have built both sorts of gatherers. Simple permanent gatherers use system tools such as SNMP probes or such as the Unix commands *vmstat, iostat, netstat* or *ps* to extract host-level statistics. More sophisticated ones address common concerns. For instance, we use a version of the Unix *df* command, modified to report information about NFS connections with gingerly probes that do not cause an NFS wait if a server is dead. We have also built very specific gatherers, launched as needed to monitor locally produced applications, such as a large infrastructure for managing scanned documents.

The Health System is written almost entirely in Java. Added to the approximately 12000 lines of Java code, a few thousand lines of C is used in data extractors. The system employs both a basic centralized repository for testing purposes, and our Bayou-based replicated repository. The Bayou port, written in java, uses MySQL as its relational database component. A designed approach to partial replication awaits implementation.

4   A Web-Based Approach: Checkup

We turn now to a description of the most extensive application we have developed to demonstrate the health architecture, a web application called *Checkup*. Checkup provides a coherent interface to a wide range of information, replacing the bag of complex software tools that a user might employ on different platforms to determine what is going on.

Figure 2 excerpts snippets from four dynamically-generated Checkup pages[1], beginning with the root page for the host AlicesHost, then progressing through the (bold-faced) links to additional views. The root page (Fig. 2, panel 1) is a top-level view of that host's overall state, wrapping the output of such standard Unix performance tools as *pstat*, *vmstat*, and *fs* into HTML presentations. This page includes links that, when invoked dynamically, produce more detailed performance analyses. For instance, following the "mounted file server" links will dynamically create root pages for those related hosts. Similarly, clicking a "per-process" link creates the page of (Fig. 2, panel 2), a table listing attributes of all running processes sorted into a user-specified order. A link from each process yields a detailed account of that process's performance and resource use: names, open local and remote files, and open network connections (not shown). Finally, one can invoke a net path analysis tool (Fig. 2, panel 3), by clicking on one of the open connections. This tool encapsulates many behavioral parameters that the user would otherwise need to supply by hand, then combines the results from both hosts into a presentation that is far more comprehensible than direct operation of the underlying UNIX tool.

This approach has several advantages over previous tools. The use of HTML hyperlinks to annotate the output of standard monitoring commands means that the Web pages define a space of information with a high degree of branching, encouraging the user to explore specific features of interest, rather than building up a general picture. The path that a user follows to a particular leaf node can contextualize the way in which the tool appearing there is used. Finally, a serendipitous advantage is that Checkup pages can readily be created with links to other web-based diagnostic, monitoring, and repair tools, extending the scope of the system with little effort.

4.1   Checkup Implementation

Checkup (see Fig. 3) is based on Java servlets, server-based objects that create Checkup pages on demand as links are followed, through the auspices of a mechanism similar to Java Server Pages [15]. To the Health Blackboard, servlets are hunters, collecting information from the blackboard to be reported back to the user. As usual, these hunter requests lead to the on-demand launch of the necessary gatherer agents, whose results are posted to satisfy the user requests. The implementation includes mechanisms for retaining (for a brief time within a

---

[1] Full-resolution color versions of all user examples are available at
  http://www.parc.xerox.com/csl/projects/systemhealth/md/examples.

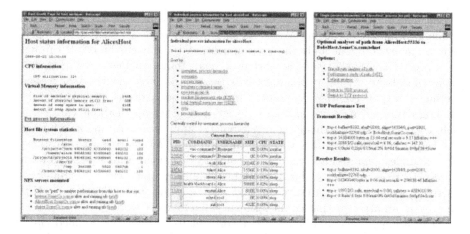

Fig. 2. A Sequence of Checkup Web Pages, Generated by Servlet Hunters.

session) time-consuming computations, such as the top-level attributes of all of a host's processes. Servlets that create related pages can use such cached information to increase responsiveness. Some pages, such as the one depicted in Fig. 2 panel 3, issue multiple requests for information provided by multiple hosts.

Servlets annotate their web pages with dynamically computed information and with hyperlinks to more detailed analyses. For example, the root page servlet checks what other systems a host depends on for its file service, and for each, either indicates that the corresponding server is not running or provides a link to the root page for that server.

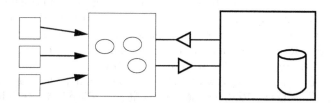

Fig. 3. Checkup Is a Servlet-Based Health Application for Standard Web Browsers

4.2   Application Helpers

We can use the same infrastructure both to evaluate and inform a specific application, by linking the application to the health system via a small "Health Helper" program.

As an example, we have built a Health-enabled variant of *emacs*, a text editor available on many platforms. We chose *emacs* because it is widely used, dependent on many components in the environment, and highly customizable through provided APIs. When *emacs* is started, a small "emacs helper" is also started to provide bidirectional communications between *emacs* and the blackboard. Acting as a gatherer, this program posts specific performance information, including details of memory and CPU usage that other applications can use to help them coexist with this large, cycle-hungry application. Acting as a hunter, the helper can supply *emacs* with information that can improve its robustness. For example, *emacs* can divert an attempt to download library code from an unresponsive server to one that is available; similarly, *emacs* can determine that there is insufficient room to save the next version in the current location, in time to prevent an attempt that could result in failure or lost work.

A Checkup page could be created to present to the user the detailed *emacs* performance information provided by the helper program. Thus, users do not have to learn a new application to monitor this application, but can use methods they are already familiar with.

5   Visualizing System State

Finally, we have been exploring the presentation of system health information through visualization tools. Visualization enables the movement of activity from the cognitive system to the perceptual system [11], making patterns and correlations in the blackboard data directly perceptible. Visualization through animation of system data [12] directly supports our intuition that patterns of significance to the user occur all the time, not only in the face of failure. Since we assert that systems health depends both on individual components and on the interactions and relationships between them, we believe a successful approach will convey these relationships as well as the more objective data. Further, we seek a presentation that minimizes the technical knowledge required to interpret it, which argues for a perceptual rather than a cognitive approach.

We have prototyped a series of simple visualizations. The *Ping Radar* visualizes the performance of a network, as seen from a particular host. In Fig. 4, columns represent hosts. Rows are snapshots at successive instants. The color of a cell estimates a round trip time to that host. The radar "beam" replaces rows cyclically at fixed intervals, giving us a picture of recent activity. Over time, patterns reveal the reliability of each connection, periods of network downtime, etc. We can see that one host appears to be down (black column), that hosts on the left are closer to us than those on the right, and that network response to distant hosts is quite variable.

Similarly, we built a *Process Map* that depicts the processes on a single host, using the position and color of data boxes to represent the memory each has consumed, their run times, and their current states. Readily perceived separations emerge between temporary small processes (simple commands), long-running processes (servers) and anomalous processes (runaway computations).

These tools are blackboard clients, permitting exploration of different ways to organize and visualize relevant information.

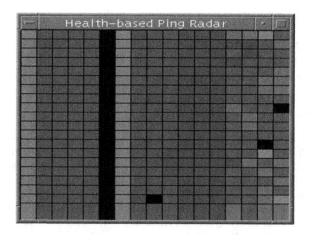

Fig. 4. The *Ping Radar* Gives a Continually Updated View of the Recent Network State

It is clear from these early explorations that the context in which information is presented is of considerable importance. Our examples so far all deal out *absolute* data: ping times, absolute memory consumption, etc. Although these readings provide valuable information, there are times when relative values, such as deviations from the norm, anomalous readings, or the differences between readings, would provide more insight.

6   Conclusions

Traditional diagnostic tools are optimized for system managers and administrators. We believe that information about the health of networked distributed systems can be of equal value to end users, if properly presented. We have the beginnings of a tool set that can tell users what is going on with their applications. We employ a blackboard architecture and multiple cooperating agents to build up information from "vertical" system views into "horizontal" slices corresponding to the end-to-end structure of user activities, combining disparate information sources into a coherent user report. This architecture is effective both for creating expressly designed health views and for augmenting the capabilities of existing applications to explain their behavior.

Basing our repository on a weakly consistent replication model enables a high degree of fault-tolerance in support of our "fail least/fail last" criterion, maximizing the availability of correct and at least partially complete information.

Checkup and our visualizers are only a first step in building health applications. Considerable work remains, particularly in the exploitation of historical information and in the ability to report coherently about application behavior. The infrastructure we have developed is a basis for further investigation.

Acknowledgments

Mark Spiteri and Karin Petersen built the Bayou-based blackboard repository. We would like to thank Anthony Joseph, Ron Frederick, and Bill Fenner for their advice and assistance, and the reviewers for many useful insights.

References

1. Brown, M. and Najork, M.: Algorithm Animation using 3D Interactive Graphics. *Proc. Sixth ACM Symp. User Interface Software and Tech.* Atlanta (1993) 93-100
2. Unicenter TNG Product Description. Computer Associates International, Inc., Islandia, NY 11788 (1998) URL: http://www.cai.com/products/unicenter
3. Distributed Management Task Force (2000) URL: http://www.dmtf.org/
4. Dourish, P.: Accounting for System Behaviour: Representation, Reflection and Resourceful Actions. In Kyng and Mattiassen (eds), *Computers and Design in Context*, Cambridge: MIT Press (1997)
5. Finkel, R.: Pulsar: An Extensible Tool for Monitoring Large UNIX Sites. *Software Practice and Experience*, 27(10) (1997) 1163-1176
6. Freeman, E., Hupfer, S., and Arnold, K.:*JavaSpaces Principles, Patterns, and Practice.* Addison-Wesley, Reading MA 01867 (1999) 344 pp. ISBN 0-201-30955-6
7. Net.Medic: Your Remedy for Poor Online Performance. Lucent Network Care, Sunnyvale, CA 94089, (2000)
   URL: http://www.lucent-networkcare.com/software/medic/datasheet/
8. Killian, T.: Processes as Files. *Proc. USENIX Summer Conf.* Salt Lake City (1984) 203-207
9. Big Brother: Monitoring and Notification for Systems and Networks. The Mac-Lawran Group., Montreal (1998) URL: http://maclawran.ca
10. Munzer, T., Hoffman, E., Claffy, K., and Fenner, B.: Visualizing the Global Topology of the MBone. *Proc. IEEE Symp. Info. Vis.* San Francisco (1996) 85-92
11. Robertson, G., Card, S., and Mackinlay, J.: The Cognitive Coprocessor Architecture for Interactive User Interfaces. *Proc. ACM SIGGRAPH Symp. on User Interface Software and Tech* (1989) 10-18
12. Robertson, G., Card, S., and Mackinlay, J.: Information Visualization using 3D Interactive Animation. *Comm. ACM*, 36(4), (1993) 56-71
13. Schoffstall, M., Fedor, M., Davin, J., and Case, J. A.: Simple Network Management Protocol. RFC 1098, SRI Network Info. Ctr., Menlo Park (1989)
14. Stasko, J., Domingue, J., Brown, M. and Price, B.: Software Visualization: Programming as a Multimedia Experience. Cambridge MA MIT Press (1998)
15. *Java Server Pages Technical Specification.* Sun Microsystems, Palo Alto, CA (2000)
   URL: http://www.javasoft.com/products/jsp/
16. Terry, D., Demers, A., Petersen, K., Spreitzer, M., Theimer, M. and Welch, B.: Session Guarantees for Weakly Consistent Replicated Data. *Proc. IEEE Intl. Conf. Parallel and Distributed Info. Syst.* Austin (1994) 140-149
17. Terry, D., Theimer, M., Petersen, K., Demers, A., Spreitzer, M., and Hauser, C.: Managing Update Conflicts in Bayou, a Weakly Connected Replicated Storage System. *Proc. ACM Symp. on Oper. Syst. Principles* (1995) 172-182

Beacon: A Hierarchical Network Topology Monitoring System Based on IP Multicast

Marcos Novaes

IBM T. J. Watson Research Center
30 Saw Mill River Rd., Hawthorne, NY USA
mnovaes@us.ibm.com

Abstract. This paper presents *Beacon,* a technique that exploits the advantages of IP multicast to provide a low overhead system for network topology monitoring. *Beacon* is a self-organizing system which builds a self similar network topology that enables the system to converge very quickly even at a high degree of scalability. Another property of the *Beacon* system is that it does not require any interdomain routing protocol for IP multicast, making its deployment possible even in subnetworks which are interconnected by domains which do not support IP multicast routing or that deploy different routing protocols. One desirable side effect of the deployment of beacon is that it acts as a self configuring IP multicast tunneling facility which provides a distributed system with fault tolerant IP multicast reacheability. A comparison of Beacon with several other routing and gateway protocols is discussed.

Keywords. Network Topology, Network Monitoring, IP Multicast, Fault Tolerant Multicast Routing

1 Introduction

The main subject of this paper is a hierarchical protocol, named *Beacon,* which was designed to provide a Topology Services facility for distributed systems. Some of these terms have been used extensively in the literature, most of the time with slightly different meaning, so it would be wise to briefly state their significance in the context of this paper:

Distributed System: A system that is comprised of a collection of nodes interconnected by a communications network; each node consisting of an instance of an operating system which is reachable via one or more network addresses. One of the most important aspects of a distributed system is the *definition of its scope,* that is, the way in which the collection of nodes is defined. Nodes may be implicitly defined by their physical placement in a particular network, such being the case with nodes which own a port to a bridge or switch. In distributed systems built for fault tolerance or for the consolidation of management tasks, the configuration of the system is done arbitrarily by the administrators of the system. In this paper, we assume an arbitrary configuration of the distributed system, but also take into account the fact that some nodes appear to the Internet Group Management Protocol (IGMP) [1] as network neighbors.

Topology Services: A facility which provides each node with the knowledge of its ability to communicate with the other nodes in the distributed system. In the context of this paper we assume that the communication property is transitive (if A can communicate

A. Ambler, S.B. Calo, and G. Kar (Eds.): DSOM 2000, LNCS 1960, pp. 169 - 180, 2000.

with B, and B can communicate with C, then A can communicate with C), and reflex-ive (if A can communicate with B then B can communicate with A). These assump-tions are generally true in the realm of layer 3 protocols, which are the subject of this paper. The literature in the field is divided between probabilistic fault tolerance sys-tems and deterministic ones. *Beacon* is a deterministic topology system, where the fail-ure of a path is actually sensed by monitoring nodes. Deterministic topology services systems rely on verification messages (which have in the literature been called *probes*, *keep alive*, *hello* or *heartbeat* messages) sent to monitoring nodes at regular intervals, and due to this fact they are sometimes referred to as *heartbeating systems*. The verifi-cation messages used in *Beacon* are (predictably) called *beacon* messages, because they are multicast messages which function as the beacon lights of a transmission tower. The transmission tower does not know who is observing the beacon lights, they are just on constantly, and are therefore always available to interested observers (low flying aircraft). This is the communication model used throughout the *Beacon* system. This communication model can be termed *observational*, because it does not employ rigorous message sequencing.

The goal of Topology Services systems is to support a high number of nodes with a minimum of overhead. Another goal is to optimize the message flow such that failures can be perceived and communicated to potentially all members of the distributed sys-tem with maximum efficiency, i.e., in a minimum amount of time. Fault tolerant dis-tributed systems usually have strict requirements for failure notifications, reaching intervals sometimes measured in milliseconds in such applications as fault tolerant multimedia streaming. The *Beacon* system which is discussed in this paper accom-plishes these goals leveraging the filtering facility present in most communication adapters which support standard IP multicast, as described in RFC 1112 [1].

2 Advantages of Utilizing IP Multicast

This section reiterates the characteristics IP multicast [1] which make this technology specially suitable in the context of distributed systems management. There are two basic aspects of IP multicast which are of special interest:

2.1   One to Many Message Propagation

As the term implies, a single IP multicast datagram can be received by a plurality of clients. The use of this facility can greatly reduce the amount of network traffic when there is a need to send a single datagram to a large number of clients. In other words, the employment of IP multicast leads to a reduction in the utilization of network band-width.

2.2   Protocol Support at the Hardware Level

The second aspect of IP multicast which makes it very attractive for distributed system management applications is the widespread hardware support for the protocol. The communication adapters which support the IGMP standard described in RFC 1112 [1] effectively filter datagrams in each network interface according to the status of IP mul-ticast group subscriptions in the host; prior to the actual receiving of these datagrams by the IP layer. This means that a multicast datagram is only received by any particular host if there is at least one process in that host which has joined that specific multicast group to which the datagram was destined on that particular interface. If there are no such subscribers, than the datagram is not received by the IP layer in the host, which means that IP multicast datagrams do not impact at all the CPU resources for a host that has no processes interested in it.

This hardware support is present in the vast majority of communication adapters in use today, and is the most attractive feature of IP multicast, and is precisely the feature that distinguishes it from similar protocols. As an example, we note that IP broadcast does have the one to many distribution property mentioned in the section above, but it has no support for isolating the set of interested receivers of a datagram. In other words, all hosts which are reachable by an IP broadcast datagram will receive it, even if there are no processes on that host interested in it.

3 Key Elements of Topology Systems

3.1 Monitoring Nodes

An important aspect of any Topology Services system is the way in which the monitoring topology is laid out. The simplest choice is to chose one distinguished node in the topology and assign to it the responsibility of monitoring all the other nodes, resulting in a monitoring topology in the shape of a star. This topology obviously cannot scale very well, as the monitoring node soon becomes overwhelmed with the overhead of receiving all the liveness messages from all other members. Another negative aspect of star monitoring topologies is that the whole system becomes unstable after the loss of the center monitoring node, and a new election has to take place and a new center node selected.

The opposite of the star topology is probably the ring topology, in which nodes are disposed linearly in an ordered stream, each node having an upstream neighbor and a downstream neighbor. Each node is then assigned the responsibility of monitoring either its upstream neighbor or its downstream neighbor. This topology is very desirable because it distributes evenly the task of receiving liveness messages from all the nodes. Nevertheless, ring topologies incur in added recovery complexity, because it is necessary to re-discover neighbors in the case of the failure of a node, and most importantly in the case that two rings have to be merged.

3.2 Leader Nodes

Another role that is commonly assigned to nodes in a Topology Services system is that of group leadership. The group leader is a distinguished node that is responsible for gathering the monitoring information supplied by the monitoring nodes, for combining the partial topology knowledge of each node and for producing a global topology view which is then sent back to all nodes. In a star network monitoring topology, the group leader is usually the center node, since it has direct access to all the topology information anyway. In ring monitoring topologies, electing a leader becomes a real problem. Since each node communicates only with its neighbors, it is necessary to devise an election scheme that traverses the ring, and that proves to be very complex in the case of failures. Therefore, most ring monitoring topologies fall back to a star topology for the purpose of leader communications. All nodes send the topology changes that they perceive (basically the death of a neighbor) to the group leader, and the leader then sends the updates back to all the other nodes. This communication pattern is not so bad, since failures are not expected to happen very often, although the leader node is likely to be overwhelmed during the bring up or shutdown procedure of a large distributed system. But the basic draw back of relying on a star topology for leadership is the cost of leader election and recovery in the case of a leader failure. The more nodes that are involved in the election process, the more complex the election will be (usually requiring the broadcasting of votes to all members). This issue is specially important when we consider the possibility of group leader failure during normal operation of the distributed Topology Services system. Not only does a new leader need to be elected, in addition the topology state must be rebuilt from scratch, usually requiring status of all monitoring nodes to be sent to the new leader, and this can result into a serious overload for the new group leader node.

3.3   Monitoring and Leader Nodes in *Beacon*

In *Beacon*, the monitoring and group leader topologies are the same, as in star topologies. That means that the monitoring nodes are also group leader nodes. That brings the advantage of having the topology information readily available in the leader nodes. The scalability problem of the star monitoring topology is dealt with by employing hierarchical star topologies, and assigning one leader for each star. The group leaders at one level of the hierarchy (also called a *tier*) will then form a star of their own and also elect a leader. Eventually the system stabilizes, having a single leader at the higher tier. This approach has the advantages associated with the simplicity of the star monitoring topology, while maintaining the capability of being able to scale to a large number of nodes.

4 The *Beacon* Protocol

Now that we have defined the scope of the problem that *Beacon* proposes to solve, and hinted at the basic improvements which can be derived from the deployment of IP multicast, and stated which key elements of the Topology System can benefit from it, let's finally explain how *Beacon* does it. In the discussion that follows, it is assumed that there is a *Beacon* process running on each node in the distributed system, which periodically sends a *beacon* on a well known IP multicast address. These addresses are also called *groups* in the context of IGMP[1], but this term makes it difficult to explain the role of nodes in *Beacon*, which forms groups of a different scope. Therefore, the word *channel* is used instead in this paper in reference to multicast addresses.

4.1   Divide and Conquer

The crudest implementation of a multicast based topology services would employ a single group leader for gathering liveness information for all other nodes. This crude implementation would no be very scalable, since the concentration of membership messages to a single node would generate an excessive load in that node. The solution is to partition the nodes into groups, which we label T_0 (tier zero) and which have distinguished leaders. Then we make it a task of the group leaders to also form a group (T_1), and elect a "leader of leaders". Once elected the leaders can forward their partial knowledge of the network topology to their group leader, which will in turn be able to combine the partial topology information collected from the other members of its group into a global view of the network topology. This global view can now reflected back to all the other members by the reverse path: the T_1 leader will send it to all the members of the T_1 group, and each member of T_1, being a leader for the members at the T_0 level, will send it to the members of its T_0 group. We now have a two tiered hierarchy.

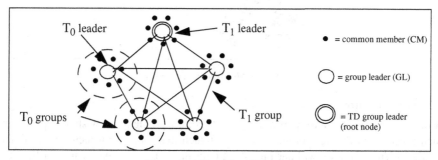

Figure 1: A Two Tiered Topology

While the two tier hierarchy is an improvement, in very large systems there will eventually be too many members at either the T_0 or T_1 levels, so it becomes necessary to employ a multi-tiered architecture. In *Beacon* this is done by limiting the number of nodes (*degree*) that a group in any given tier may have. The maximum number of members per group is presently read from a configuration parameter, but an interesting extension of this work would be to determine this limit adaptively from the state of the system. The lower the group degree, the more tiers the resulting group hierarchy will have. The illustration below shows a fully populated beaconing network of degree 5.

The numbering of tiers in *Beacon* is done as follows:

a. The number 0 is distinguished, and derived from an implicit rule: network neighborhood. An instance of a T_0 group is a group of nodes which are network neighbors.

b. The root node R belongs to T_D, where D is the *depth* of the spanning tree rooted at R. The root node is elected using the election protocols detailed in the sections that follow.

c. Each intermediary tier is numbered by subtracting from D the number of edges required to reach the tier departing from R.

We note that in the figure below that the root (indicated by a double circle) is a leader of a Tier 2 group, since the depth of the tree is two. It is also a leader of a tier 1 group, which is represented by the smaller star attached to it. And it is also a leader of a T_0 group, as indicated by the cloud of common members which orbit it. The root node is the only node that may be a group leader at three tier levels. The remaining nodes can be group leaders at most at two tier levels.

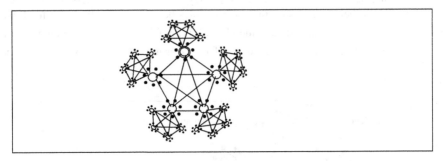

Figure 2: A Multi-tiered Topology

4.2 Tier Zero Groups

Now that we have a tool for partitioning the Topology problem, we wish to deploy it in such a way such that it exploits IP multicast such that the system incurs in minimal overhead. Therefore, we will partition the groups in a way that makes sense for IP multicast.

The group forming process outlined above can be seen as a bottom up procedure that forms a tree. We now need to define how each node is placed as a member of a T_0 group. In *Beacon,* Tier-0 is chosen to be formed of the nodes that are network neighbors in terms of IP multicast. That is, two nodes belong to the same T_0 group if and only if they can exchange IP multicast messages with the Time to Live (TTL) parameter set to 1, i.e., without having the messages traverse any router. This means that the set of possible nodes in each T_0 group is implicitly determined by the way in which the network is configured.

4.3 The *Beacon* Double-Channelled Communication Model

Now that we have established an implicit rule that divides the nodes in T_0 groups, we can now study how to gather topology information at the T_0 level and elect a leader that will forward this information up in the tree. Since we have now reduced the number of nodes that participate in the protocol to a relatively small number, we can now use the simple procedure of having each node send beacon messages to a well known multicast address (channel), and have all nodes monitor each other. But this is clearly a waste of CPU utilization, since it is only necessary for one distinguished node to act as a monitoring agent and report the topology information to the next tier. The other nodes would be wasting CPU cycles receiving beacon messages, since they are not reporting it to the other tiers.

A better approach is to first have the group election, select a group leader and then have the leader be the only node in each T_0 group which actually joins the multicast channel that receives beacon messages. Since this channel is used by the other members to send beacons to the group leader, we label it the Group Leader (GL) channel. By allowing the other members to filter out the beacon messages, we allowed them to save CPU cycles. Nevertheless it is now necessary for all the other nodes in the T_0 group to monitor the health of the group leader, in order to make sure that they have a representative at the higher tiers. Therefore, all nodes in a T_0 group which are not the group leader join a second multicast channel, which we label the Common Member (CM) channel. This channel is used by the group leader to send its own beacon (GL beacon) to the other members and also any updates to the network topology. We are now exploiting the many to one distribution of IP multicast, and thus saving the network bandwidth. Therefore, by using two multicast channels we have reduced the CPU utilization and the network bandwidth used in the protocol. We still have kept the protocol simple enough so that it can be written with a few lines of multithreaded code.

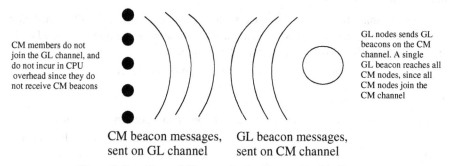

CM members do not join the GL channel, and do not incur in CPU overhead since they do not receive CM beacons

GL nodes sends GL beacons on the CM channel. A single GL beacon reaches all CM nodes, since all CM nodes join the CM channel

CM beacon messages, sent on GL channel   GL beacon messages, sent on CM channel

Figure 3: Two Channel Communication Pattern Used in Beacon

4.4   Election Protocols and Group Formation

Now that we have decided on the communication model that we want to have between the Common Members and their Group Leader, we now need to specify a procedure with which the members of the group can unanimously arrive at such an arrangement.This is done with a very simple election routine that is done at initialization time on every node, and also every time that a CM has not received messages from its GL within a configurable period of time. This election procedure is similar to the root node election used in the IEEE 802.1d link layer protocol. Another similar variation is used in IGMP version 2. The election procedure that follows is an adaptation of this simple procedure to the double channelled communication model:

1. CM members *always* emit periodic beacons on the GL channel. This is elegantly done by having a separate thread dedicated to this task. The beacon messages are used both to report liveness status, and also for election purposes. Beacon, messages are *always* sent with TTL=1, such that they are contained within the T_0 domain.

2. When entering the election procedure, the CM node will join both the CM and the GL multicast channel. This is the only time that a CM node has to join the GL channel.

3. After joining both the CM and GL channels, the CM member waits for a period of time *e,* during which the CM member records the messages that it receives. The outcome of the election will be determined by the beacon messages received within the period *e*. Note that there is no new message type associated with the election, and also that there is no specific election protocol. The election is simply done by having each electing member observe the traffic in both multicast groups for a period of time and then to deduct the state of the system using the following rules:

4. If the CM member receives a beacon in the CM channel, this means that this T_0 domain already has a leader. The CM node than leaves the GL channel and there after

runs in CM mode. This mode basically consists of continuing to beacon periodically to the GL channel and of monitoring the health of the group leader by receiving messages in the CM channel. The CM node will only exit this state if it fails to receive beacons from the GL member, in which case it will re-initiate the election, going back to step 2 above.

5. In case that the CM node does not receive any beacons in the GL channel, then it will assume that the T_0 domain was undergoing election and that the node that should be chosen as the group leader is the node that sent a beacon message from the interface with the lowest IP address.

6. The node that contains the interface with the lowest IP address will immediately start sending beacons on the CM channel. It will also assume the role of Group Leader and will initiate the search for higher numbered tiers.

7. All other nodes resume operation as CM nodes, proceeding as in 4.

4.5   Interdomain Elections

As a result of the tier zero elections, a collection of independent T_0 groups has been formed. The leaders of these groups will now search for each other, with the objective of attaching their T_0 domain to a higher tiered group. This operation will produce beaconing structures which involve multiple T_0 groups. In the discussion below, these beaconing structures are called *domains*.

In order to unite the separate T_0 domains, we will utilize the initial assumption that the list of all IP addresses for all nodes in the system is available from the distributed system management facility. Each GL member which has been elected at tier N will make a search for other tiers consulting this list. We can optimize this search according to the way that we chose T_0. If T_0 corresponds to a subnetwork, it is possible to send a subnet directed broadcast querying for the leader of each T_0 domain. In the following discussion we label the node performing the search S, and the nodes responding to search queries R:

1. The search space consists of all the nodes in the list. The first step is for the S to delete from the search space the nodes that it knows to be located within its own T_0 domain (since these nodes were already discovered by the initial election).

2. An IP address is now selected from the list, giving priority to the addresses which can be reached with the least number of routing hops. A point to point connection to the node is attempted. If it fails, we choose another address and retry. This is repeated until a responsive node U is found.

3. Node S now sends a query to U, requesting its status (CM, GL or undefined).

4. If R replies that its status is undefined, it means that it is undergoing election. Then S will wait for a period of time e and expect to be informed of the new status of the node and proceed as below.

5. If the R replies that it has CM status, then U will direct S to the group leader of the highest numbered tier of its domain. Then S will terminate the connection to U, and contact the GL member named in the reply and proceed as below.

6. In the remaining possibility, S eventually contacts a node which is a GL in another T_0 domain. We label this node L. The two nodes then exchange the topology information related to their domains, that is, the list of nodes which they have already discovered. This information exchange is all that is needed for the election. Again, the results are determined by an implicit rule derived from the network addresses

7. Using a rule similar to the one used in the T_0 election, the lowest numbered IP address is used to determine if L or S wins this election. The leader of the domain which contains the lowest IP address (and which is by definition a tier leader) will take no action. The other node, will attempt to join the other domain, by selecting a group leader in the other domain that is the closest to it, according to some metrics such as the number of IP hops (easily obtainable, given we have the IP address), and also that has not reached its maximum degree of members. This closest leader is labeled the attachment point, A. The join procedure is initiated by the joining node, which sends a join message to A.

8. If A can accommodate a new member, then the join proceeds and the joining node assumes the role of a CM node in the higher level tier, being responsible to sending beacon messages to its leader and also being responsible to monitoring the leaders health.

9. If the joining node cannot find any tier leader that it can join, then it will chose a CM node that is already attached to the tree (preferably the CM that is closest to the joining node according to the number of routing hops) and make a request to initiate a new lower level tier. The chosen CM node will then become the GL of the newly formed tier.

Once a searching node finds another domain with a lower IP address, it will join this domain and stop the search. Therefore, each domain will have only one node which continues the search procedure, which is the leader of the highest numbered tier, or *root* node. A root node will stop to search for other members if it detects that it has found all possible T_0 domains, i.e., if any of the following conditions apply:

a. It looses an election to a lower addressed domain, and becomes a CM node in a tier of the winning domain.

b. A node detects that all T_0 domains are represented in it domain. This condition also means that this node is the root of a spanning tree that connects all T_0 domains in the system. The search can then stop because all T_0 domains have been found.

4.6 Node Monitoring in Tiers above T_0

Now that we have glued together the T_0 domains into a hierarchical structure, we need an infrastructure with which members of a higher tiered domain can monitor themselves. Again, the number of nodes in each tier was limited to a tractable number, and so even the least sophisticated procedures would work. Basically it is just necessary for the CM members of a domain to send beacons to their SL, and the SL has to respond with a beacon to all its members. A straight point to point approach would work, but why develop new code when we already solved the problem for the T_0 case? By virtue of being networks neighbors, the T_0 members are able to utilize the two IP multicast channels and save in resource utilization and programming complexity. The logic of the topology structure is self similar, and therefore the code should be reusable. All that is needed is for each member of a domain to establish an IP multicast tunnel to each other member, and make all group leaders run an instance of a standard IP multicast routing protocol, such as DVMRP, or PIM.

We can now run the very same procedure used for T_0 monitoring, and have the same savings, although they will come for different reasons. In the T_0 level, having only the GL node join the GL channel means that all other nodes will filter out the unwanted beacons. In levels above T_0, that are not network neighbors, not having any other node join the GL channel means that the tunneling nodes that have no members for that group will send prune messages back to the source of the beacons, having the effect that beacons are only effectively transmitted to the GL, the only node that is actually interested in them. On the other hand, all nodes join the CM channel and act as tunneling endpoints, and therefore the beacon messages from the GL are routed to all nodes, and the network bandwidth is thus spared by virtue of the one to many transmission pattern.

4.7   Recovering from Failures of Links in the Hierarchical Tree

The most important benefit of the *Beacon* hierarchy is probably the simplicity of the recovery procedure in the case of the loss of nodes. This capability should be studied in the context of the recovery of spanning trees, and may have many applications in layer 2 protocols. Each star in the beacon hierarchy is a separate recovery domain. A failure of a node in a specific domain is dealt with only by the members of that domain, and will make minimal impact to the rest of the tree. For example, consider the failure of the root node. In non hierarchical systems the entire tree would have to be recalculated. Now, let's examine how this is handled in *Beacon*. The root is the leader of the highest tiered domain, T_D. Its failure is directly sensed by all members of T_D, which will then promptly elect a new leader. The tree now has a new root, and no new edges had to be created or recalculated. Nevertheless, the failed leader at the T_D level was also a leader of T_{D-1}, by virtue of the hierarchical construction of the tree. Therefore all members that were connected to the tree via the T_{D-1} domain of the failed leader are now disconnected. The members of T_{D-1} domain also sense the failure of their leader, and will then elect a new leader which in turn will re-attach the disconnected subtree at some point of the main tree. The important aspect here is that neither the main tree nor the disconnected subtree were dissolved, and do not have to be recalculated. Lastly, we note that the members of the T_0 domain that contained the failed leader are still disconnected. They perform a similar procedure, electing a new leader which then re-connects the T_0 domain back to the main tree.

The recovery capability of *Beacon* preserves the calculated leaders and edges even in the face of an arbitrary number of failures. Each failure will cause at most tree disconnected domains, which elect new leaders and attempt to re-connect to a higher priority tree. Eventually all the trees are reconnected with minimal disruption or loss of calculated leader and edges.

5 Relationship to Previous Work

Curiously, the majority of the literature in Topology Services systems does not come from the area of distributed system management, but rather from the area of communications. We mentioned previously the similarities with the IEEE 802.1d protocol [10], which is a layer 2 protocol. We can also find some heartbeat capability added to standard routing protocols, such as RIP [11], OSPF[4], DVMRP[5], PIM-SM[12]. The heartbeating capability of these protocols is usually an added feature to the main protocol, provided to give the protocol the capability of coping with router failures. The main difference is that whereas these protocols require the configuration of redundant routers such that there are alternative paths to route around failures, *Beacon* configures the smallest number of routers such that there is a tree that spans all members. *Beacon* copes with failures by having all nodes in the distributed system acting as stand by routers, and dynamically configures nodes as router nodes as needed. This contrasts with the traditional approach of configuring redundant routers because it saves on the amount of route announcements and router traffic overhead. If we were to configure every single node of a distributed system as a router node, than the amount of router announcements and router traffic would be enormous. Also, it would create looped networks, which would cause most dynamic protocols to converge slowly or not at all.

The proposal in *Beacon* is to precisely abstract out the Topology functionality from the routing domain and put it back in the distributed system management domain, where it can be better controlled. It was mentioned earlier that a distributed system may be comprised of several nodes taken from subsections of network domains that run different layer 2 and layer 3 protocols. *Beacon* allows for the uniform control of the topology parameters (such as the heartbeat interval, which determines the speed with which failures are sensed). Just because we have a system that contains a few nodes from a huge Ethernet LAN that need to be monitored very closely because they are video stream servers, it does not follow that all the routers in this large LAN should be setting their *hello* messages to 10msec intervals. Actually, most protocols recommend intervals that range from 30 sec. to 90 sec., far exceeding the range for most critical applications.

Beacon also relates to border gateway protocols such as BGP[6] and BGMP[9], in the sense that it offers interdomain multicast routing. Actually, *Beacon* could be deployed as an interdomain multicast routing protocol in conjunction with either DVMRP, MOSPF or PIM-SM or PIM-DM. But *Beacon* does not deal with individual multicast groups, it just establishes tunnels and lets another protocol, such as DVMRP deal with the group subscriptions. Finally, *Beacon* is hierarchical, and while an attempt at providing hierarchy for interdomain routing protocols was made with BGP Federations [8], the approaches are very different. The *Beacon* hierarchy is self configurable and self similar, very different from the proposed, statically configured flat hierarchies proposed in [8].

References

1. S. Deering, "Host Extensions for IP Multicasting, IETF RFC 1112, 1989
2. A. Ballardie, "Core Based Trees Multicast Routing Architecture", IETF RFC, 1997
3. J. Moy, "Multicast Extensions to OSPF", IETF RFC 1584, 1994
4. J. Moy, "OSPF Version 2", IETF RFC 2328, 1998
5. D. Waitzman, C. Partridge, S. Deering, "Distance Vector Multicast Routing Protocol", IETF RFC 1075, 1988
6. Y. Rekhter, "A Border Gateway Protocol 4 (BGP-4)", IETF RFC 1771, 1995
7. A. S. Thyagarajan, S. E. Deering, "Hierarchical Distance-Vector Multicast Routing for the MBone", ACM SIGCOMM, 1995
8. P. Traina, "Autonomous System Confederations for BGP", IETF RFC 1996
9. D. Thaler, D. Estrin, D. Meyer, "Border Gateway Multicast Protocol (BGMP): Protocol Specification", IETF Internet Draft, 2000
10. R. Perlman, "Interconnections: Bridges, Routers, Switches and Internetworking Procols", 2nd. ed., Addison-Wesley, 1999
11. G. Malkin, "RIP Version 2", IETF RFC 1058, June 1998
12. D. Estrin et. al, "Protocol Independent Multicast - Sparse Mode (PIM-SM): Protocol Specification", IETF RFC 2362, June 1998

Managing Concurrent QoS Assured Multicast Sessions Using a Programmable Network Architecture

Radu State, Emmanuel Nataf, and Olivier Festor

LORIA - INRIA Lorraine - Université de Nancy II
615 rue du Jardin Botanique
F-54602 Villers-les-Nancy Cedex, France

Abstract. In this paper we address the management of concurrent multicast sessions with security and QoS guarantees. Their main feature is a high degree of change in terms of membership, implying the necessity for fast reconfiguration and provisioning. We will approach the management problem using techniques developed in the context of virtual private networks, adapted to the high dynamicity that we are confronted with. We propose a framework for the management of such networks by integrating the management and the control plane, using the programmable and active networks paradigms developed within the research community. We apply the framework to the management of residential user TV multicast, where an ATM based access network supports the delivery of TV content to all clients having subscribed to a DVPN.

Keywords: Active Network, P1520, VPN, multicast.

1 Introduction

The advent of broadband technologies for the local loop, combined with the deregulation of this part of the network in most European countries, fosters the deployment of new services to the end user. One of these services is the multicasting of digital TV channels to all subscribed residential customers. Such a service requires both a dedicated signalling plane (e.g. for channel selection by the end-user) and a high performance management plane for provisioning, monitoring and flow management.

The project aims at providing a management framework for concurrent QoS assured multicast sessions in the backbone (here a metropolitan area network) to provision the local loop. Each TV channel is conceptually defined as a multicast tree which has the characteristics of a VPN, ie. Closed User Group, security and QoS guarantees. The support of multicast facility within VPNs is required in order to optimize the use of network resources. Since the required multicast trees are strongly dynamic, their configuration within the traditional VPN management time-scale is no more viable. Our work proposes an integration of the management and signalling planes using programmable and active technologies in order to cope with this strong variability and dynamics.

A. Ambler, S.B. Calo, and G. Kar (Eds.): DSOM 2000, LNCS 1960, pp. 181–192, 2000.
© Springer-Verlag Berlin Heidelberg 2000

To present the major components of our architecture, the remainder of the paper is organized as follows. Section 2 gives a definition of Dynamic Virtual Private Networks, illustrates the target backbone to be managed and motivates the need for a programmable architecture in order to enable in-time management. Section 3 and 4 detail the components of our management architecture. In section 5 we detail the distributed software architecture. Section 6 summarizes the work done in other projects which has been partially reused in our approach. Finally a short conclusion is given and future work is outlined.

2   Dynamic Virtual Private Networks

Dynamic Virtual Private Networks are from a logical point of view virtual private networks which encapsulate multicast trees whose topology may change very frequently depending on user interactions (join, leave a channel). These DVPN rely on Service Level Agreements (SLA) established between individual end-user subscribers and the service provider as well as on SLAs between the latter and content providers. The service provider may itself rely on a transport provider with specific SLAs. At the lowest level, the network is composed of ATM switches. At the backbone edges, customers access the dynamic virtual network through Service Access Points (SAP). The physical infrastructure uses the Asymmetric Digital Subscriber Line (ADSL) technology.

Each TV channel is modeled as a DVPN. This choice is appropriate since common features which are particular to VPNs (e.g: Closed User Group, Security and QoS guarantees) are of a crucial importance in the context of TV residential broadcast.

Traditional Management time-scale for VPN provision is not appropriate for such DVPNs. In fact, the topology change of each tree must occur in a couple of milliseconds whenever a customer zaps from one channel to another. Moreover, this configuration task must be performed in parallel enabling multiple customers to change channels simultaneously. Since generic network layer support is unlikely to support such requirements, part of the management tasks must be integrated with the signalling facility into a programmable infrastructure. Time constraints are one motivation, information sharing provides a second one. Both camps benefit: like in active networking, signalling protocols may benefit from information provided by the management framework (like topology or link states), and on the other hand, a management framework may benefit from information provided by the value added services signalling plane (e.g. actual number of customers who are looking at a given channel).

3   The Management Architecture

In order to enable management of those DVPNs, we have chosen to combine the programmable network approach and active technology, as detailed in this section.

As illustrated in figure 1, the management architecture is built within a CORBA DPE on top of P1520 [3,2] abstract switch interfaces. The entire architecture is composed of five elements.

The main component is the DVPN Tree Manager. This entity is responsible for the configuration, extension and reduction of DVPN trees on the backbone. It maintains a view of the topology of the physical network as well as a logical one for each each DVPN currently in activity. This entity offers an API to the Customer Service Access Point (second component) through which all channel setup/change requests are received, acknowledged (check that the user is allowed to join a given DVPN) and performed (issue an expansion request to the DVPN for the given SAP and the given channel.

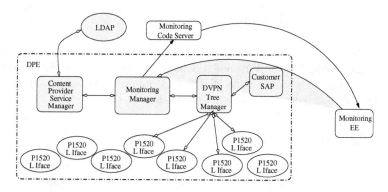

Fig. 1. The Management Architecture Building Blocks

The third component is a monitoring entity. This management entity provides facilities for deployment of monitoring code for both the Service Provider and the Content Providers. Related to the monitoring entity is the Monitoring EE, an Execution Environment for active code. This EE is availaible in various nodes of the network, especially on each edge node as one special End-User. This EE is used by the Monitoring manager to deploy service specific monitoring code for one or multiple content provider service parameters. This monitoring code is developed by the network management staff according to SLAs. The resulting code can be dynamically downloaded and controlled by the Monitoring Manager in accordance with the service management entity described below. The functionality that we achieve with this approach corresponds to a on-the-fly construction of RMON Mibs. For usual traffic monitoring, the monitoring manager uses SNMP-based agents located in the switches, at least in the first release of the management environment. The monitoring manager relies on a monitoring code server offering a set of monitoring functions which can be deployed to the probe EEs onto the network.

The last component is a service level manager called Content Provider Service Manager, responsible for both DVPN setup and collection of data that must be made available to the content provider (mainly based on what is specified in the Service Level Agreement. This element mainly relies on the monitoring manager to gather information from the network. A second information source is the Customer SAP through which coverage related data can be obtained. The service level reports are made avalaible to the content provider through an LDAP directory server.

All those components are currently under development using the Java technology and a CORBA DPE at the programmable level. The Execution Environment and active code is an extension of the ANTS Toolkit [12] running on a Linux box for probes.

4   DVPN Tree Manager

The DVPN Tree Manager is the core of the management platform.We will first describe the static information model, that is the structure of the information needed to maintain a view on the DVPN and public network properties. Afterwards we will continue with the functional aspects of its activity.

4.1   Information Model

The information needed in order to perform the management is twofold, since two different layers of abstraction can be put into evidence. The first one concerns the public network used to deliver the DVPN service. The underlying public network is modeled using a slightly modified information model of the one introduced in [10]. The extensions to that model concern the support of multicast connections at a layer trail and respectively subnetwork connection layer. At the DVPN level, new entities are introduced and linked to the supporting elements from the public network layer view (see figure 2). For the sake of clarity we keep the public network part of the information model quite simple and generic. Specific ATM related entities are obtained by a specialisation of the generic classes. For instance :

- the LayerTrail class can be derived in order to stand for a ATM SVC. It has several ATM related properties in terms of traffic descriptors, QoS service class and parameters.
- By a specialisation of the Subnetwork class, one can model physical ATM switches. However, this class can also model a collection of ATM switches performing as one global switch.
- SNC (Subnetwork Connection) represents physical cross-connects done in the switches, but also communication across a collection of switches, considered to jointly form a Subnetwork entity. It is worth noting that the SNC object has several destination endpoints and one designated source endpoint in order to perform at switch level the point to multipoint cross-connections.

- nwCTP models connection endpoints. In the case of a ATM residential backbone, it is mapped to a triplet (port, VPI, VCI) representing the interface
  and the connection identifiers for the particular connection.
- Link represents the connectivity at the physical layer.
- LinkTP models the termination point of a Link object and is mapped to the
  particular switch interface used.

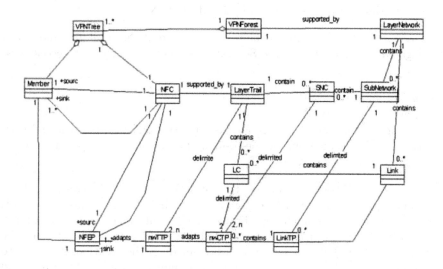

Fig. 2. DVPN Information Model

The information model for VPNs, that we extended for reuse in the context
of DVPN, has been introduced in [8]. We enhance the model in order to be able
to:

- model a collection of DVPNs as seen by a global view. The class VPNForest
  is introduced for this purpose.
- represent the Multicast Tree that corresponds to a particular DVPN. At the
  DVPN level, such a tree is a collection of network flow endpoints (NFEP) and
  the traffic from one source NFEP to the remaining ones. This traffic is modeled
  using the network flow connection class (NFC) and represents the video data
  delivered to the end users. The NFC is a one-to-multipoint connection in order
  to model a multicast tree. It is characterized by a series of QoS parameters.
  If MPEG format is used, these parameters can be for instance:

- the I, B and respectively P frame rates.
- MPEG/ATM encapsulation parameters : PCR aware/PCR unaware, MPEG2 to AAL5 transport stream packet adaptation issues, average per flow jitter.

The active code based monitoring is performed on this type of objects. The content provider is able to check if SLAs parameters are satisfied by deploying application specific code. For instance, it can simulate client behaviour and measure the performance of the provided service.

- A DVPN member corresponds to a a network address, used by a particular residential user, and that is subscribed to DVPNs.

The dynamic reconfiguration of the tree is initiated at this level first and down-propagated to the network level. Several causes can trigger such a reconfiguration.

- users that join a DVPN,
- users leaving a DVPN,
- Management initiated actions. For instance, stationary parts of the multicast tree, can be merged onto a one-to-many permanent connection in order to facilitate management.

For instance, if one user desires to join a DVPN, its SAP asks the DVPN Tree manager to add a branch to the particular NFC. At the network level, this operation is translated in adding a branch to the trail supporting the NFC. Since a LayerTrail is build from interconnected SNC and LC, these objects must be created/modified. The usual operation flow involves the addition of branch to an existent SNC object that is already used to multicast the DVPN traffic, and the necessary LC and SNC objects are created in order to deliver it to the users's access point. In case of users leaving a DVPN, there might be the case that whole areas of the multicast tree need to be teared down in order to release bandwidth resources.

5   The Distributed Software Architecture

The software architecture, illustrated in figure 3, is composed of two main parts. The first one regards the clients objects used to access specific interfaces on the service provider system, whereby the second one is concerned with the implementation specific issues in the service provider domain.

5.1   The Client Side

The only object visible to the outside of the platform is the VPNMemberFactory object. Clients will use the Corba Naming Service to be able to reference this objet. Its main responsibilities are related to the authentification of users and the processing of their requests.

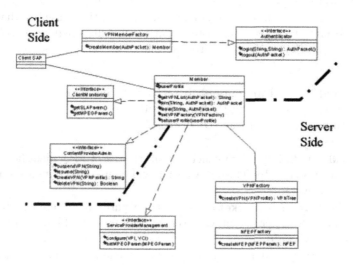

Fig. 3. Distributed Software Architecture

Authentication and Service Access Interfaces. The authentication part is dealt with by implementing the `Authentificator` interface which consists of two methods:

- `login(String, String)`. Using a centralized database where the client's profile is stored, and the provided userid and password, users are offered access to the service. In case of a successful login operation, an `AuthPacket` is returned to the user. This return value will be further used by the user for all his successive operations. In this phase of the project it represents the user's cresidentials in terms of subscribed VPNs, and specific Service Level Agreements.
- `logout(AuthPacket)` will simply logout the user and clean up resources.

The returned `AuthPacket` packet will be used to authenticate every operation. This is an additional security precaution since a malicious user could try a masquerade and obtain a reference to already instantiated `Member` objects to access and use the service at someone else's charges.

The main functionality of the `VPNMemberFactory` object is to create `Member` objects. This is done using the `createMember(AuthPacket auth)` method, which creates the `Member` object to be used by the user to join/leave particular VPNs. Additionally, users can use this object in order to

- get statistics concerning service usage and preliminary billing information.

– configurate simple policies in terms of VPN access. Access to a particular VPN could be prohibited for a delimited period of time based on the rating of its information content. For instance, movies having a high degree of violence could be blocked in order to protect young watchers.

Operational and Management Interfaces. The `Member` class implements one operational interface and three management interfaces. The operational interface is used to join and leave DVPNs.

– `getVPNList(AuthPacket)` which returns the list of VPN identifiers, that the user can subscribe to. This information will be used to configure some of the user's hardware.
– `join(String, AuthPacket)` connects the particular Member object used by a user having cresidentials given by the auth AuthPacket to the VPN identified by vpnId.
– `leave(String, AuthPacket)` leaves the current VPN, that the client is connected to.

The management interfaces provide for several types of management initiatives. The residential user, the content provider and the service provider are allowed to perform management actions. The first one, called `ClientMonitoring` permits the monitoring of SLA established parameters at the client side. Such type of parameters are for instance:

– parameters related to zapping. A simple SLA could specify that connecting to a particular VPN, as a result of a user zapping action, should not take more that 250 ms.
– parameters related to the MPEG stream like for instance the received I, B and P frame rates, and associated measurements.

A Content Provider is a special type of a user having more facilities in terms of rights to create and delete VPNs and perform restricted custom management on its VPNs. This is done through a private interface `ContentProviderAdmin` implemented by the `Member` class. A Content Provider can access this interface only if its `AuthPacket` packet encodes these type of cresidentials. The methods included in this interface permit to create/delete and manage a particular VPN by a content provider. Administrative issues (like client subscribing and unsubscribing operations) but also global MPEG/ATM related performance measures can be addressed, if necessary.

5.2   The Server Side

The `Member class` exports a management interface to the Service Provider to be used in the end user management and monitoring.

VPN End User Configuration and Monitoring. The Service Provider can access the `ServiceProviderManagement` interface to perform management operations on to `Member` objects. This interface is accessed by the Service Provider in order to perform application specific management configuration and monitoring. In the case of MPEG over AAL5 video delivery, such management pertains to the:

- configure hardware specific parameters (eg. interface,VPI/VCI).
- setting the value of MPEG transport stream packets, encoded in one AAL5-SDU. One transport stream packet is 188 bytes such that if only one packet stream is encoded in one AAL5-SDU, one needs 5 ATM cells for the their transport. If more transport stream packets are encoded in one larger AAL5-SDU, bandwidth is used more efficiently. The default value is 2. Several choices are possible based on jitter requirements. For instance, a large value will provide for efficient bandwidth utilisation requiring though a higher jitter.
- Setting the MPEG stream rate in conformance with the ATM Layer traffic description. Several mechanisms for this purpose are introduced in [1,11].

This interface is also used by the Service Provider in order to verify that SLA's in terms of residential users MPEG quality is assured. In order to allow for more flexibility for this purpose the Service Provider can deploy active code at the user's site, using this interface.

New DVPNs are created through a a `VPNFactory` object. The method `createVPN` is called on behalf of a `Member` object when a Content Provider deploys a new DVPN. The argument of this method is an object of the type `VPNProfile` encapsulating VPN related SLAs. In our case, they concern the MPEG-2 end-to-multi-end properties. These properties will be used to create the corresponding `NFC` object. The resulted `VPNTree` is registered with the `VPNForest` entity, and its identifier is returned to the user. A `VPNTree` object corresponds to the administrative view of a DVPNs. It includes methods to add/remove members, and provides for general membership related statistics (average time of connection, average number of users, max user number). Specific MPEG-2 related parameters, like the ones mentioned above, are accessed via the NFC object attached to a `VPNTree`.

Dynamic VPN Membership Management. Let us consider the case of one client joining a DVPN, illustrated in figure 4 .

In this case the method add(client-member) is called on the corresponding `VPNTree` object. This results in the creation of a `NFEP` object, its addition to the associated `NFC` and its configuration according to the `NFC` parameters. Next, using the associated `Member` object, the `nwTTP` to be used is determined. A client profile includes a `nwTTP` determined by the access equipment and the triplet(port, VPI/VCI) used on the latter. The `LayerTrail` object used to support the `NFC` object is determined and required to add the `nwTTP` object. Since one can map a `LayerTrail` object onto a one-to-multipoint `SNC` objet, a similar Subnetwork

Connection Management structure to the one introduced in [7] can be used. As soon as the `Member` object has been added to a DVPN, a packet of the type AuthPacket is returned. This packet will be used by the methods implemented through the `ContentProviderManagement` interface, in order to authentificate management actions performed by the Content Provider. The global view of currently existent DVPN is available to the network manager via the `VPNForest` object.

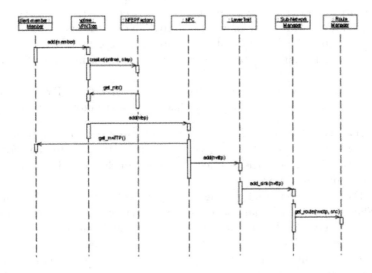

Fig. 4. DVPN Join Operations

6   Related Work

From the architectural framework point of view, our approach relies on the APIs defined in the IEEE PIN project and at the Columbia University. Currently we base our specifications on the ATM Switch Resource Abstractions APIs. Concerning VPNs, the Genesis project [5] proposes a distributed network operating system based on those APIs as well as a spawning architecture for their setup. Their objectives are quite different from the ones defined in our framework. The Genesis kernel could be used in our framework to spawn an initial DVPN. The idea of combining programmable and active technology has already been proposed in the mobiware framework [6,4], where the programmable framework was used to build the signalling plane, and the active technology permitted to inject code to adapt the data plane to wireless customer-tailored specific QoS conditions. The principle developed in our approach is conceptually similar but applied to the management plane for the active technology part. The use of active technology for management has been proposed several times over the last year. Most of these approaches deal with standard delegation, mostly in conjunction with a legacy SNMP environment. This is done for instance in the SmartPackets approach [9] where active packets are used to delegate monitoring and access to SNMP variables. Our approach for the active technology part is different from those approaches in the way that the active code sent to monitoring Execution Environments is not dependent on standard SNMP, but provides application specific flow monitoring facilities to instrument the edge probes.

7   Conclusion and Future Work

In this paper, we have presented a framework which combines both a programmable paradigm and the use of active technology for the monitoring of dynamic virtual private networks. The use of a programmable network architecture is motivated by the need to combine both application level signalling and network level management.

At the programmable network level we have implemented a DVPN model and a prototype configuration manager. This manager is fully integrated in the DPE, enabling thus efficient information exchange with the signalling facility.

Using active technology for dynamic monitoring represents in our framework an extension to management by delegation. In fact, using this technology enables deployment of management functions that are adapted to a given type of application, gathering the semantics of the data flows. This is very interesting for making service management more efficient and user-friendly.

The current approach is limited to multicast trees based on a single source for each VPN and multiple data sinks. One future direction of this work will be to extend the approach to multi-source multicast groups.

References

1. Audiovisual multimedia services: Video on demand specifications 1.1. *The ATM Forum*, March 1997.
2. C.M. Adam, A.A. Lazar, and M. Nandikesan. ATM Switch Resource Abstractions, March 1999. IEEE/WG P1520/TS/ATM-017 Working Document.
3. C.M. Adam, A.A. Lazar, and M. Nandikesan. Switch abstractions for designing open interfaces, March 1999. IEEE/WG P1520/TS/ATM-016 Working Document.
4. O. Angin, A.T. Campbell, M.E. Kounavis, and R.F. Liao. The Mobiware Toolkit: Programmable Support for Adaptive Mobile Networking. *IEEE Personal Communications Mag., Special Issue on Adapting to Network and Client Variability*, 5(4):32–44, August 1998.

5. A. Campbell, H.G. De Meer, M.E. Kounavis, K. Miki, J. Vicente, and D.A. Villela. The Genesis Kernel: A Virtual Network Operating System for Spawning Network Architectures. In *2nd Int'l Conf. on Open Architectures and Network Programming*, N.Y., May 1999.

6. A.T. Campbell, M.E. Kounavis, and R.F. Liao. Programmable Mobile Networks. *Computer Networks and ISDN Systems, Computer Networks*, 31, April 1999.

7. J. P Gaspoz. *Object Oriented Method and Architecture for Virtual Private Network Service Management*. PhD thesis, École Polytechnique Fédérale de Lausanne, 1996.

8. E.C. Kim, C.S. Hong, and J.G. Song. The multi-layer vpn management architecture. In *Proc. NOMS 1999*, pages 187–199, 1999.

9. B. Schwartz, W. Zhou, A.W. Jackson, W.T. Strayer, D. Rockwell, and C. Partridge. Smart Packets for Active Networks. In *Proc. OpenArch '99*, March 1999.

10. R. State, E. Nataf, and O. Festor. Poster session: A Java based Implementation of a Network Level Information Model for the ATM/Frame Relay Interconnection. In *Proc. NOMS 2000*, April 2000.

11. C. Tryfonas and A. Varma. MPEG-2 Transport over ATM Networks. *IEEE Communications Survey*, 2(4):24–32, 1999.

12. D.J. Wetherall, J.V. Guttag, and D.L. Tennenhouse. ANTS: A Toolkit for Building and Dynamically Deploying Network Protocols,. In *Proc. IEEE OpenArch'98*, 1998.

Providing Global Management Functions for ATM Networks

Bernhard Thurm[1] and Hajo R. Wiltfang[2]

[1] Institute of Telematics, University of Karlsruhe
D-76128 Karlsruhe, Germany
thurm@telematik.informatik.uni-karlsruhe.de
[2] Deutsche Telekom AG, IP Backbone Engineering, D-48014 Muenster, Germany
Hajo.Wiltfang@telekom.de

Abstract. Modern computer networks with their increasing bandwidth and their ability of QoS support require adequate management solutions. For ATM networks, a management system especially needs global functions combining the local management functionality distributed over all individual network devices to network-wide functions in order to manage e.g. end-to-end connections and their QoS aspects. In this context, the following paper presents an approach to ATM network management which introduces a function-oriented architecture. Based on this architecture, the functionality of two developed management entities, the ATM Network Monitor and the PVC Manager, which are both designed for important network-wide management functions like topology discovery, connection management, and QoS monitoring are described in detail.
Keywords: Distributed management, Network-wide management functions, ATM networks, ATM management architecture and applications

1 Introduction

Computer-based communication has become a more and more central aspect for modern society. Besides traditional data communication such as email and file transfer, forthcoming applications like multimedia and tele-conferencing increase the need for integrated networks supporting Quality-of-Service (QoS) on a per connection basis. An important example in this area is the ATM (Asynchronous Transfer Mode) technology which provides virtual connections (VCs) based on asynchronous cells and a dedicated QoS associated with each connection.

For integrated networks such as ATM, the number and quality of additionally offered services comes along with a rising complexity of the network technology in general and of the required network devices in detail. This rising complexity leads to significantly increased requirements on an adequate network management. Especially the QoS aspect together with the connection-oriented communication scheme demands for totally new management functions. In traditional networks, management functions are mainly concentrated on individual network devices, e.g. configuration or monitoring of a router. In case of ATM, new *network-wide* management functions are needed in order to combine the local functions distributed over all individual network devices with the aim of controlling and monitoring the global functionality of this integrated network

A. Ambler, S.B. Calo, and G. Kar (Eds.): DSOM 2000, LNCS 1960, pp. 193–204, 2000.
© Springer-Verlag Berlin Heidelberg 2000

technology. For example, the management driven set-up of a VC between two given end points throughout the network requires management element functions that are related to more than one individual network device. Existing approaches to ATM management do either not focus on network-wide functions or they rely on special kinds of interfaces to the management of individual devices. For instance, the VIVID system from Newbridge [6] uses the proprietary CPSS (Control Packet Switching System) protocol on a dedicated management VC for providing network-wide functions whereas the xbind architecture [4] is built on top of CORBA [3] which requires a CORBA interface at each device.

To overcome the described demand for network-wide functions, this paper presents a management approach which is based on standardized and widely deployed management interfaces and information as well as a developed function-oriented management architecture. Because of its orientation towards the required functions this architecture is specially designed for supporting network-wide ATM management. All details and characteristics of the so-called *FuMA architecture (Function-oriented Management architecture for ATM networks)* are described in Sect. 2. Based on FuMA, our work concentrates on the development of network-wide functions for some key areas of ATM network management: topology discovery, VC management, and QoS monitoring. As one main result of our development and implementation efforts, Sect. 3 presents the *ATM Network Monitor (ANEMON)*, a management entity providing functions for all areas mentioned above. In a next step, Sect. 4 describes the developed management application *PVC Manager* which offers an easy-to-use Web-based interface for all management functions developed so far. In order to show the practical usability of our management functions, Sect. 5 presents the results of some simulations done in a very flexible and scalable simulation environment especially designed for ATM network management simulations. Finally, Sect. 6 summarizes the key points of this paper and gives some ideas for future work.

2   Function-Oriented Management Architecture for ATM Networks

The Function-oriented Management architecture for ATM networks (FuMA) [12] has been developed as management framework in order to provide three abstraction levels for ATM management functions. Each of the three levels focuses on different types of management functions, starting with the simple functions distributed over the devices, continuing with the more complex network-wide functions, and ending up with user interface functions. In the following, the three levels element management, management middleware, and management applications will be described in detail (see Fig. 1).

At the base level, the *element management* deals with the management of individual ATM devices. Hiding specific characteristics, the element management offers a well-defined management interface for each ATM device. Our specification of this interface includes the type of management protocol as well as the information provided. In detail, the commonly accepted Simple Network Management Protocol (SNMP) is used for all management communication within FuMA and the management information assumed for each ATM device is based on the standardized ATM-MIB [8]. Moreover, integration

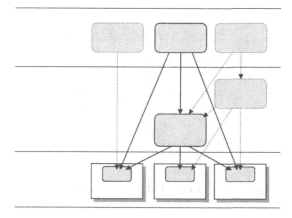

Fig. 1. FuMA Architecture

of devices which do not support the (whole) ATM-MIB can be achieved by using a special kind of proxy agent [7].

At the second level, the *management middleware* provides network-wide functions which are related to a whole network and thus, to more than only one ATM device. These functions combine the local functions distributed over the individual devices and therefore, are far more complex than the simple element management. The main advantage of the middleware results from the concentration of complex network-wide functions in some modular entities which can be used by multiple other management entities in parallel. Thus, for different management applications, there is no need to implement the middleware functionality by themselves. For example, the ATM Network Monitor ANEMON (see Fig. 1) offers network-wide management functions for topology discovery, VC management, and QoS monitoring, as detailed in Sect. 3.

At the third level of FuMA, *management applications* are located. They typically provide a (graphical) user interface for all functions offered by middleware and element management. Thus, applications directly interact with human users wanting to manage ATM networks. As one example, Fig. 1 shows the developed application PVC Manager which will be presented in Sect. 4.

3   ATM Network Monitor

The management middleware entity *ATM Network Monitor (ANEMON)* has been designed in order to provide network-wide management functionality for the tasks topology discovery, VC management, and QoS monitoring in ATM networks. Based on the modular concept of FuMA middleware, each of these functions can be seen as an independent functional block with its own dedicated MIB describing the implemented functionality. The ANEMON combines three functional blocks within one middleware entity, but the modular structure still remains as shown by Fig. 2.

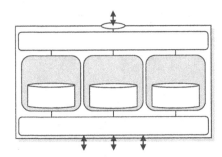

Fig. 2. Architecture of the ATM Network Monitor (ANEMON)

Besides the three functional modules, which will be described in the following sections, the architecture of the ANEMON comprises two further modules (see Fig. 2). The SNMP agent handles the SNMP-based communication with management applications or other middleware entities (see Fig. 1) in order to provide the implemented management functionality of the ANEMON. The amount of management information offered by this agent is given by the superset of all functional module MIBs which will be introduced in the following sections. At the interface to the ATM devices, the SNMP manager of the ANEMON performs the communication needed for obtaining all required information from the devices.

3.1   Module for Topology Discovery

Topology information has a very fundamental character for network-wide management functions on which this paper concentrates. In order to identify the ATM devices involved in a network-wide function the performing management entity essentially needs information on the underlying network topology, i.e. which devices are connected to each other. Therefore, the first and most fundamental functionality developed for the ANEMON was the topology discovery which is located within the topology module (see Fig. 2). The management information of this module is specified by the developed *ATM-TOPOLOGY-MIB* [12,13] which includes information for configuring and controlling the topology discovery as well as providing the resulting topology data.

The algorithm implemented for *topology discovery* is based on a passive concept, i.e. topology information already distributed between ATM devices is collected and combined to the overall topology. In detail, neighborhood information contained in the *atmInterfaceConfTable* of the ATM-MIB [8] is used by the algorithm. Starting at one given point, the algorithm computes the overall topology step-by-step by continuously evaluating the distributed neighborhood information. In each step, all neighbors are determined for the evaluated ATM device and then, the search is applied to each neighbor in a depth-first search manner. The developed discovery algorithm terminates at each detected end system or if the configured maximum search depth is reached. More details on this algorithm can be found in [12,13].

3.2   Module for VC Management

For the management of virtual connections (VCs), two types of network-wide functions can be distinguished, tracing of existing VCs (for any kind of VC) and management (i.e. set-up and disconnect) of permanent VCs (PVCs). For both areas, our VC module (see Fig. 2) provides the basic set of required management functions. Again, the management information for this module is specified by a developed MIB, our *ATM-VCC-MIB* [9,12].

The developed algorithm for *VC tracing* is very simple because it can take advantage of the topology information already provided by the topology module (see Sect. 3.1). Based on this information and a given VC starting point, the algorithm computes the trace by continuously evaluating VC switching information on the next ATM device in the topology [12]. The result mainly consists of a set of connection links (VCLs, virtual channel links) which are stored in a special data table of the ATM-VCC-MIB.

The functions developed for *PVC management* are also based on our topology information, however, they are much more complex. Primarily, this is caused by the fundamental concept for virtual ATM connections which requires a virtual channel connection (VCC) to consist of a series of virtual channel links (VCLs). Each VCL is identified by a combination of VPI (Virtual Path Identifier) and VCI (Virtual Channel Identifier) unique for the underlying physical link. The concatenation of these virtual links is obtained by cross-connecting them at the intermediate ATM devices. Consequently, the management of a whole VCC can only be achieved by considering all distributed devices. As an example, we will illustrate the different steps and algorithms involved in the set-up process of a permanent VCC.

First of all, an 'appropriate' path between two end points has to be calculated. Considering certain constraints (unambiguousness, absence of cycles, optimality) together with the similarity of ATM networks and undirected distance graphs, we developed a solution based on Dijkstras algorithm for the *single source shortest paths* problem [2]. Necessary modifications primarily were related to the different types of ATM devices (end systems perform no switching) and parallel physical links [9,12].

Step two concerns the selection of a VPI/VCI combination to identify a connection. Locally, the only management information available from the ILMI-MIB and the ATM-MIB [1,8] define certain intervals for the identifiers of virtual channel connections. However, searching through a column of the *atmVclTable* containing the VPI and VCI values in use can slow down the establishment process if lots of virtual links are configured throughout the network since it has to be performed for each device on the path. Globally, i.e. considering the whole end-to-end span, it is desirable to identify single virtual links of a connection consistently. Therefore, the best trade-off between an arbitrary and a time-consuming full uniform identification is to find a pair of VPIs and VCIs which is usable for the two end points of a connection while for the inner links changes are made only when necessary [9].

In a third step, all virtual links, cross-connects, and traffic parameters have to be created which can be achieved by accessing the *atmVclTable*, the *atmVcCrossConnectTable*, and the *atmTrafficDescrParamTable* of the ATM-MIB located on each device.

3.3   Module for QoS Monitoring

Finally, the QoS module of the ANEMON is designed to provide functions for monitoring QoS parameters in ATM networks. So far, two functional areas are addressed by our QoS module, the monitoring of statistics for physical links within the topology (aggregated link-based QoS) and QoS monitoring for selected VCs which focuses on each VCL of a monitored VC (detailed VCL-based QoS). The management information for both functional groups is specified by our developed *ATM-QOS-MIB* [10,12].

The *monitoring of physical links* is based on the interface information of the MIB-II [5] (e.g., the octets sent and received at each ATM interface of involved devices) and thus, offers statistics to the overall usage of an ATM network [10]. By providing various configuration options within our ATM-QOS-MIB, a very flexible network performance monitoring can be obtained this way.

The objective of the developed *VCL-related QoS monitoring functions* is to provide detailed diagnostics when one end point of a VC detects problems with the end-to-end QoS. In that case, the only way to identify the responsible device within the network requires measurements on each VCL of the connection and thus, on each ATM device involved. To that aim, we developed a set of monitoring functions [12] which are based on the VC tracing described in the previous section. The requesting user only has to specify one end point of the VC and the QoS module automatically starts to monitor the requested parameters on each link. The resulting measurements are then provided in a special data table of the ATM-QOS-MIB.

4   PVC Manager

Management communication within FuMA is generally based on SNMP. Thus, our primary goal in designing the *PVC Manager* was to implement a FuMA management application providing a Web-based interface to management middleware and element management, but requiring no knowledge of SNMP itself [9,12].

Focusing on permanent virtual connections (PVCs), the PVC Manager integrates a comprehensive amount of functionality concerning the areas configuration, performance, and fault management. Moreover, the implementation as Java applet guarantees the flexibility and extensibility necessary in today's heterogeneous networks.

As shown in Fig. 3, the architecture of the PVC Manager is basically composed of four distinct modules: topology discovery and display, connection management, device information, and QoS monitoring. While the first two modules are explained more detailed in the following sections, the functionality of the latter two is given very shortly: Basic management information about single ATM devices (e.g. the *ifTable* of the MIB-II) is provided by the *device information* module, whereas the *QoS monitoring* module is responsible for triggering the monitoring process in the ATM-QOS-MIB as well as for acquiring and displaying measurements periodically [10].

4.1   Topology Discovery and Display

The module *topology discovery and display* is essential for the graphical user interface of the application. After launching the applet, the ANEMON is contacted and used to

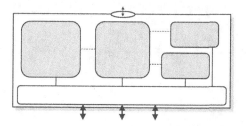

Fig. 3. Modules of the PVC Manager

generate the graphical representation of the network topology by reading the ATM-TOPOLOGY-MIB (see Sect. 3.1). Each device is classified as end system, VC/VP switch, or VP switch (recognizable by the usage of different icons) while the capacity of single physical links is outlined by the thickness of the corresponding line (see Fig. 4). Concentrating on management functionality itself, no algorithm for the 'perfect' placement of all topology elements was included. Instead, the basic layout can be modified by the user (the graphical representation shown in Fig. 4 is the result of manual adjustments as well). Finally, as all modules rely on the process of topology discovery, the parameters of the topology module of the ANEMON are fully configurable, for example, starting point of the search, update interval, and search depth.

4.2 Connection Management

The module *connection management* implements the interface to the VC related functions of the element management as well as the access to the VC management module of the ANEMON. Because of its complexity, the module itself was split into three different sub-modules VC configuration, VC handling, and VC information.

Initially, VC configuration provides an efficient interface to the *atmVcl/VplTable* and the *atmVc/VpCrossConnectTable* of the ATM-MIB which simplifies the set-up of virtual links and cross-connects by the usage of graphic dialogues instead of inconvenient SNMP operations. In addition to useful filtering capabilities (e.g. only existing, non-terminating VCLs are candidates for cross-connecting), row-creation, activation, and configuration of traffic parameters in the *atmTrafficDescrParamTable* are not visible and performed as one single step. Furthermore, by relying on the ATM-VCC-MIB of the ANEMON, the PVC Manager allows the set-up of complete permanent virtual connections. This includes the selection of source and destination (or the whole transmission path) of a connection by simple mouse-clicks as well as the configuration of all other significant parameters (see Fig. 5). However, the procedure of actual VC set-up is entirely left to the ANEMON by using the ATM-VCC-MIB (see Sect. 3.2).

Supplementary functions are located in sub-modules two and three, VC handling and VC information, which primarily are responsible for provisioning functions concerned with the deletion of virtual links, cross-connects, and connections along with the display of important managed objects of the ATM-MIB. Another feature included

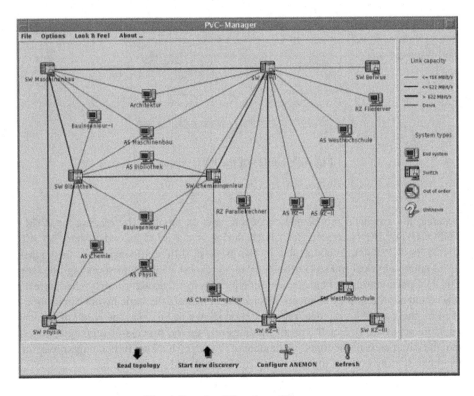

Fig. 4. Result of Topology Discovery

is the tracing of existing connections, which is based on the corresponding function of the ATM-VCC-MIB (see Sect. 3.2).

5   Evaluation in a Simulated Environment

In most of the tests performed, we used a simulation environment which has been specifically designed for evaluating ATM management functions [11]. The whole simulation environment is based on so-called *virtual ATM devices* consisting of a software management agent which operates on internal data only and thus, is able to simulate the behavior of a physical ATM device in a realistic way. The main advantage of this simulation approach compared with a hardware testbed is its scalability because small software agents each simulating one physical device can be placed in almost any amount within a simulation set-up. Another aspect is the very short time necessary to construct and simulate different scenarios.

The simulation scenario chosen for this presentation is based on a network topology which has been derived from the topology of an existing network, the ATM backbone of the University of Karlsruhe [11,12]. Focusing on ATM devices only, this backbone network consists of 9 switches and 13 end systems connected by physical links which

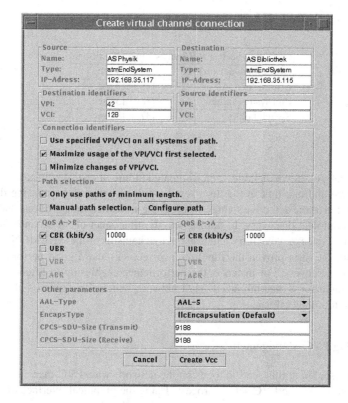

Fig. 5. Dialogue for the Set-Up of a VCC

run at 622 Mbps between switches and 155 Mbps between switches and end systems (see Fig. 4). On top of that scenario, we tested the ANEMON together with the PVC Manager as described in the following two sections.

5.1 Testing Topology Discovery

In this test, the PVC Manager was used in order to configure the topology discovery of the ANEMON (e.g. starting point), to initiate the discovery process, and to display the resulting topology. Fig. 4 shows the test result as provided by the ANEMON and displayed by the PVC Manager (after some manual re-arrangement, see Sect. 4.1) when all devices in the scenario are operational.

Our performance measurements [11,12] show that all 22 devices and 38 links of the topology are discovered correctly in approximately 4 seconds, even if some of the physical links are not in operational state (see Table 1). Though the effect of a failure of a whole device depends on several factors, obviously all devices within the topology are completely recognized only if there is at least one "bypassing" link, i.e. if the network does not fall apart into isolated sections by the removal of the defective device and its

Table 1. Performance of the Topology Discovery

Discovery Scenario	Devices	Links	Time Span
Operational network	22	38	4 sec.
Broken link	22	38	4 sec.
End system failure	22	38	38 sec.
Switch failure	21	37	38 sec.

adjacent links. As shown in Table 1, the simulated failure of one switch leads to the loss of one device and one link in the discovered topology because the missing device is only connected to the defective switch, whereas the simulated failure of one end system leads to a complete discovery. In both cases, the time span of 38 seconds is mainly determined by the extensive retry and timeout mechanisms of the ANEMON.

Over all, the tests proved the functional correctness and a good performance for the topology discovery. For instance, the performance enables up to fifteen discovery cycles a minute for a medium sized ATM network which is absolutely sufficient for rather static topology information.

5.2   Testing PVC Management

The tests presented in this section focus on the network-wide PVC management functionality, especially the PVC set-up process. Using the graphical interface of the PVC Manager, the appropriate inputs are made and then, the set-up process is triggered at the ANEMON. If the set-up is successfully completed, the PVC Manager displays a new window (see Fig. 6) containing the resulting VC data provided by the ANEMON. Moreover, the path of the new VC is highlighted in the topology view (see Fig. 4).

Table 2. Results of PVC Set-Up Tests

No.	Function	Description	Time Span
1.	Creation of a VCC	QoS set to CBR (10 Mbps) for both directions, 4 hops (3 physical links), pre-configured transmission path and VPI/VCI	8 sec.
2.	Creation of a VCC	Same configuration as 1. but with automatic path calculation	8 sec.
3.	Creation of a VCC	Same configuration as 2. but automatic VPI/VCI selection	9 sec.

As Table 2 shows, the whole set-up process takes only 8 seconds in this test. Additionally, another test turned out that despite the simulated network is of considerable size, the time needed for path calculation can be neglected since the test took 8 seconds

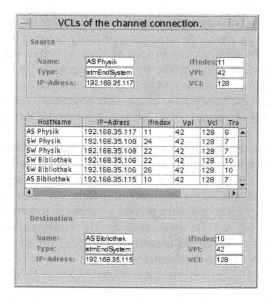

Fig. 6. Result of PVC Set-Up

for a pre-configured path as well. However, if the connection identifiers are chosen automatically (taking about 9 sec.) the whole set-up process is highly depending on how many VCs (i.e. VPIs and VCIs) are already in use.

To summarize, the PVC set-up tests proved the functionality and performance of our developed management entities. A PVC set-up time of about 8 seconds is a very good result for our medium sized network environment considering the amount of SNMP operations that have to be carried out in order to enter the desired values in the local MIBs distributed over all involved devices.

6   Conclusion and Future Work

In this paper, we presented our approach for ATM network management which mainly focuses on the important network-wide management functions required for ATM. The framework of our approach is given by the developed Function-oriented Management architecture for ATM networks (FuMA) which classifies management functions according to their functionality and view. Two of the components developed for our framework have been described in detail: Firstly, the ATM Network Monitor (ANEMON) which provides management functions for the important areas topology discovery, VC management, and QoS monitoring, and secondly, the PVC Manager which implements a graphical user interface to all functions of the ANEMON as well as to a lot of enhanced management functions of individual ATM devices. In the performed tests, both presented management components proved their functionality and showed a good performance using a specially designed simulation environment. To summarize, our approach provides a framework and some of the most essential functions for network-wide ATM management. Focusing on standardized and commonly accepted management interfaces only, the FuMA approach can be seen as a first important step towards global ATM management which is able to operate in a heterogeneous environment including ATM devices from different vendors. Moreover, the presented concept of the FuMA architecture, especially the management middleware, is generally applicable to management in distributed environments when global or network-wide management functions are required.

Future work primarily is concentrated on applying the developed and successfully tested management entities to a physical network. To that aim, we will cooperate closely with the University Computing Center of Karlsruhe in order to deploy our solutions in the real ATM backbone of the university. Currently, our main activities focus on adapting our topology discovery to the proprietary management interface of the ATM devices in that network and afterwards, continuously obtaining statistical data on the daily traffic by using the developed monitoring functions [10]. Besides practical tests, future work also concerns the development of additional or extended functions for our two entities as well as some more management entities for our FuMA framework.

References

1. The ATM-Forum. Integrated Local Management Interface (ILMI) Specification Version 4.0. af-ilmi-0065.000, September 1996.
2. E. W. Dijkstra. *A Discipline of Programming*. Prentice-Hall, 1976.
3. Object Management Group. Common Object Request Broker Architecture. Rev. 2.0, July 1995.
4. A. A. Lazar, S. K. Bhonsle, and K.-S. Lim. A Binding Architecture for Multimedia Networks. *Journal of Parallel and Distributed Computing*, 30(2):204–216, November 1995.
5. K. McCloghrie and M. Rose. Management Information Base for Network Management of TCP/IP-based internets: MIB-II. RFC 1213, March 1991.
6. Newbridge. MainStreetXpress Network Management Release 2.0. General Information Book, 1998.
7. H. Ritter, F. Fock, and H. R. Wiltfang. A Flexible Management Gateway for ATM-based Networks. Internal Report No. 20/98, University of Karlsruhe, Department of Computer Science, September 1998.
8. K. Tesink. Definitions of Managed Objects for ATM Management. RFC 2515, February 1999.
9. B. Thurm. Web-based Management of ATM Connections. Master's thesis, Institute of Telematics, University of Karlsruhe, Germany, February 1999.
10. B. Thurm and H. R. Wiltfang. Link-based Performance Monitoring of ATM Networks. In *Proceedings of the 25th Conference of Local Computer Networks (LCN)*, Tampa, Florida, USA, 8. - 10. November 2000.
11. B. Thurm and H. R. Wiltfang. Simulating ATM Network Management using Virtual Devices. In *Proceedings of the 2000 IEEE / IFIP Network Operations and Management Symposium*, Honolulu, Hawaii, USA, 10. - 14. April 2000.
12. H. R. Wiltfang. *Function-oriented Management of heterogeneous ATM Networks*. PhD thesis, Institute of Telematics, University of Karlsruhe, Germany, February 1999.
13. H. R. Wiltfang and F. Fock. Topology Discovery for ATM Networks. In *Proceedings of the 9th IEEE Workshop on Local and Metropolitan Area Networks*, Banff, Canada, 17. - 20. May 1998.

MetaNet: A Distributed Network Management System for Information Delivery and Planning with Quality of Service Guarantees[1]

Narayanan Natarajan, Abdelhakim Hafid[2], Ben Falchuk, Arunendu Roy,
Jorge Pastor, and Young Kim

Telcordia Technologies, Inc., 331 Newman Springs Road, Red Bank, NJ, 07701, USA
ahafid@research.telcordia.com

Abstract. *MetaNet* is a network management system that provides end-to-end communication services with QoS and communication planning support for mission critical applications. It supports immediate and advance resource reservation, and provides information on immediate and future available communication capacity taking into account network configuration and traffic. MetaNet supports these services over multiple heterogeneous networks. The MetaNet consists of a number of *PoP*s (Point of Presence); each PoP is associated with one or more communication networks. PoPs realize the functionality within their networks and coordinate with each other to realize the end-to-end functionality. Associated with each network is an *Adapter* that hides from the PoP the network technology specific QoS model and mechanisms.

1 Introduction

Information delivery and planning support for mission-critical applications with QoS guarantees can be achieved by a three-tier system architecture [1]. The top layer, the *policy layer,* maps mission objectives into resource policies (such as user priorities and resource limits) that guide network resource allocation among competing users. The middle layer, the *information channel control layer,* receives requests for information channels from applications and assigns to the channels QoS, priority, start time, and stop time. It optimizes network resource utilization by making use of policies set by the policy layer and current/future network states reported by the bottom layer, the *resource management layer.* The resource management layer manages network resources spanning multiple heterogeneous networks (ATM, IP, Wireless, etc.), provisions QoS for information channels taking

---

[1]This work was sponsored by the Defense Advanced Research Projects Agency. The content of the information does not necessarily reflect the position or the policy of the Government and no official endorsement should be inferred. The U.S. Government is authorized to reproduce and distribute reprints for Government purposes notwithstanding any copyright notation thereon.
[2] Contact author.

A. Ambler, S.B.Calo, and G. Kar (Eds.): DSOM 2000, LNCS 1960, pp. 205 - 217, 2000.
© Springer-Verlag Berlin Heidelberg 2000

into account resource availability, channel priority, and schedule. It is worth noting that a basic QoS support, by the underlying networks, is assumed; for example, in IP networks diff-serv [2], int-serv [3], and/or MPLS [4] support is assumed.

This paper presents the design and implementation of a distributed system, called *MetaNet* that provides the resource management layer functions outlined above. The client of MetaNet is called Agile Information Controller (AIC). The MetaNet is distinguished from other current-day QoS management systems in the following aspects:

- It provides end-to-end communication services, called *MetaNet channels*, over multiple heterogeneous networks. The end-to-end services may be IP or ATM services.
- It supports multi-dimensional QoS including bandwidth, delay, and loss.
- When it requests a channel, the AIC can specify a QoS region (using a range or a set for each QoS parameter) and MetaNet provisions QoS within the region. This minimizes MetaNet-AIC interactions. This approach is more flexible than primary-alternate QoS descriptors supported in ATM signaling [5].
- MetaNet supports priority management. It preempts lower priority channels to accommodate the QoS of higher priority channels. This is an important requirement in mission critical applications that is not supported in commercial products.
- AIC can request channels to be setup either immediately or at a future time. This feature is again important in mission critical applications. As an additional feature, MetaNet allows AIC to specify a start time range, and selects the start time that meets the channel QoS requirements. This capability is not supported in commercial products
- To enable AIC to assign "realistic" QoS for channels, MetaNet provides an abstract end-to-end view of the static and dynamic (i.e., over a specific time interval) availability of network resources. By "end-to-end", it is meant that MetaNet hides details of intranetwork and internetwork topology. Thus, MetaNet supports queries like "What is the maximum capacity possible for a potential channel between Host A and Host B during the time interval t1 to t2?". Both individual channel constraints and aggregate network constraints (e.g., maximum capacity possible between two hosts) queries are supported. Static resource constraints are determined by network configuration, such as link capacity limits, and multipoint capabilities. Dynamic constraints are determined by network load. This feature is rather unique to MetaNet and is fundamental to service planning in mission oriented environments.

Design of MetaNet supporting the above features requires solutions to the following challenging problems:

- To provision QoS for channels, MetaNet needs to mimic the QoS provisioning scheme used by each underlying network; this is required if the underlying network technology does not allow route enforcement (e.g., ATM does not allow one to specify an explicit route for SVCs).
- MetaNet needs to incorporate an algorithm that efficiently computes the network state as a function of potential channel end points, channel lifetime, and priority. This algorithm should be scalable to large numbers of networks and

channels. A short description of the algorithm (QoS provisioning algorithm) is presented in Section 3; a more detailed description can be found in [6].

- Another major challenge is the definition of the abstract end-to-end network view that MetaNet provides to AIC and an algorithm for processing queries on end-to-end resource availability. The AIC-MetaNet interface is described in Section 2 and the algorithm is introduced in section 3; a more detailed description of the algorithm can be found in [6].

- To cope with multiple heterogeneous networks, the MetaNet architecture should clearly separate technology specific part from technology independent part. A two-tier architecture adhering to this principle is described in Section 2.

- *Internetwork QoS signaling:* A signaling protocol across diverse interconnected networks is needed to realize the end-to-end functionality of MetaNet. More specifically, such a protocol should allow for the support of all features of MetaNet in terms of priority management, advance reservation of resources, and current/future network state computations. None of the existing signaling protocols for ATM and IP networks satisfies these requirements. In Section 2, we briefly present a QoS signaling protocol that supports these requirements.

2   MetaNet Functional Architecture

One of the key challenges in MetaNet design is that MetaNet should be able to support seamlessly network heterogeneity and scalability. To satisfy these requirements, we developed a two-tier MetaNet system architecture (see Figure 1).

Points of Presence (PoP)
Each PoP is associated with one or more networks. Each PoP realizes its portion of the MetaNet functionality within the associated network(s) and coordinates with other PoPs to realize the end-to-end functionality across all networks under the purview of MetaNet. Inter-PoP interactions are supported by an internetwork QoS signaling protocol supporting multi-dimensional QoS, priority based channel preemption, and resource constraints queries and responses.

Adapter
Each adapter is associated with one network. The adapter hides the network technology (e.g., ATM, IP, tactical networks) and QoS mechanisms, including QoS model and signaling protocols, from the PoP while providing a number of primitives that allow the PoP to access the functionality provided by the associated network in a generic manner.

2.1   MetaNet Channel Management and QoS Model

MetaNet provides functions for the management of point-to-point and point-to-multipoint channels. This includes channel setup, release, modify QoS, and modify topology (i.e., add/remove sinks) functions. MetaNet supports both immediate and advance reservation of resources for MetaNet channels.

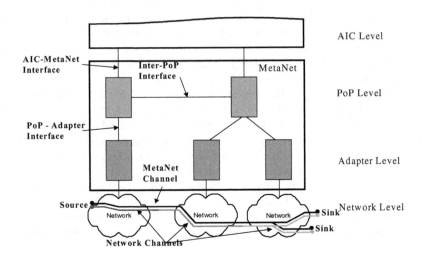

Fig. 1. MetaNet Architecture

When defining the QoS model to be used in MetaNet, the challenge we faced is that such a model should allow for multi-dimensional QoS specification (a) using generic QoS metrics that are not unique to any network technology; (b) for any type of traffic (e.g., video, WWW); and (c) in terms of QoS regions to minimize the interactions/negotiation between MetaNet and its clients. (a) and (b) are required in order to support the heterogeneity in terms of networks and traffic assumed by MetaNet. (c) is required to provide AIC with the flexibility in specifying QoS (e.g., bandwidth ranges) via a single request and thus, optimizing the response time.

The QoS parameters that are supported by the MetaNet QoS model are *bandwidth*, *delay* and *loss ratio*. These QoS parameters are defined for an uni-directional MetaNet channel originating from the source network interface to the sink network interface traversing the communication networks under the MetaNet. For each MetaNet channel, AIC specifies a range for bandwidth, delay and loss ratio parameters. Of the three QoS parameters, the bandwidth parameter is mandatory while the other two QoS parameters may be left unspecified in a request to indicate a "don't care" or "not applicable" value.

Three classes are identified for the bandwidth parameter specification: constant bandwidth, variable bandwidth, and adaptive bandwidth.

Constant Bandwidth
Constant bandwidth is specified for traffic that is delay sensitive and has constant requirements over time. It is specified as a range that consists of the worst acceptable value (minimum range value) and the preferred value (maximum range value).

Variable Bandwidth

Variable bandwidth is specified for traffic that is delay sensitive and has variable requirements over time (e.g., MPEG-2 Video). The bandwidth specification consists of a set of the peak bandwidth and the average bandwidth. This is very useful when AIC wants to setup a channel to transmit video, for example; if AIC wants (in decreasing order) to send video as a MJPEG, MPEG-2, or H.261 depending on the availability of network resources, then it may specify three pairs of (peak bandwidth, average bandwidth) in the request (a pair per video traffic type). Upon receipt of this request, MetaNet will try to setup a MetaNet channel to transmit MJPEG; if this is not possible, it will try to setup a channel to transmit MPEG-2; if this is not possible, it will try to setup a channel for H.261. This type of negotiation is not supported by any of the existing QoS models. For instance, ATM UNI [5] allows for the specification of only ATM traffic descriptor & the Minimum acceptable ATM traffic descriptor or ATM traffic descriptor & alternative ATM traffic descriptor. In the video example described above, two ATM SETUP requests may be required. Using MetaNet, one needs only one request.

Adaptive Bandwidth

Adaptive bandwidth is specified for traffic that is neither delay nor loss sensitive and that requires only a minimum bandwidth. For this class, bandwidth specification consists of the worst acceptable value (minimum range value) and the preferred value (maximum range value).

2.2   Abstract Network View and Resource Constraint Queries

To perform its resource planning function, the AIC needs an abstract view of end-to-end network resource availability. For example, AIC needs information about the maximum capacity possible, over a given time interval, for several potential MetaNet channels; e.g., "What is the maximum capacity possible for a potential channel between Host A and Host B during the time interval t1 to t2 without any preemption and with preemption of all channels with priority smaller than priority p1?". Using this information along with the mission objectives, AIC should be able to determine the channels (and their attributes in terms of QoS, priority, start time and stop time) to be setup.

Providing AIC with information, for example, about the maximum available capacity of potential MetaNet channels is not enough. AIC also needs to know the impact on the capacity available to one channel if another channel is setup with a specific capacity, i.e., bandwidth tradeoffs among potential channels. The challenge is the definition of the necessary information that MetaNet should provide to AIC without incurring an unacceptable overhead and/or violating the MetaNet philosophy of hiding the underlying network details from AIC.

Our solution, that satisfies these requirements, is based on the concept of *MetaNet Link*, defined below, and providing the maximum channel capacity and the maximum aggregated capacity of MetaNet links of interest including information on their dependencies.

MetaNet Link
It is defined as a virtual link between two hosts. It is an end-to-end concept and it does not represent the internal structure of networks under MetaNet. If MetaNet is used to manage only WAN connectivity, routers in LANs can be considered as hosts.

Maximum Channel Capacity
Maximum channel capacity of a MetaNet Link is the maximum capacity that can be requested for a single MetaNet channel between the hosts on the two ends of the link.

Maximum Aggregate Capacity
Maximum aggregate capacity of a MetaNet Link is a ceiling on the sum of the capacities of all potential MetaNet channels that can be setup between the hosts on the two ends of the link.

Dependency
It is a binary relationship defined on MetaNet Links as follows. Two MetaNet Links are dependent on each other if a potential MetaNet channel over one link may compete for common resources with a potential MetaNet channel over the other link. Determination of this relationship is based on intra-network and inter-network resource allocation (including routing) algorithms and policies used within MetaNet.

Using these concepts, MetaNet supports queries on specific MetaNet links, including their static and dynamic capacity over a specific time interval and at a specific level of priority, and dependency relationships.

2.3   Inter-Pop Signaling

PoPs realize the functionality locally within their associated communication networks, and interact/collaborate with each other to realize the end-to-end MetaNet functionality. The set of these interactions is called *inter-PoP signaling*.

In defining the inter-PoP signaling, a number of challenging issues arise: (a) the protocol should not be technology dependent; and (b) the protocol should allow for transmission of information necessary to support MetaNet's unique features such as advance reservation, priority management and static/dynamic resource availability computations. The existing signaling protocols do not tackle any of these issues. In this section, we briefly outline the salient features of our inter-PoP signaling by describing the inter-PoP interactions involved in setting up a MetaNet channel.

When a PoP receives the setup request (from AIC), it determines the internetwork route for the channel (source routing), determines the availability of its local network resources, possibly after determining the set of lower-priority channels to preempt, "conditionally" commits the local resources to the new channel, and propagates the setup request to the next hop-PoP, including the route information. Each PoP repeats this step.

The PoPs involved in a MetaNet channel setup effectively commit the network resources for the channel only when all PoPs determine that resources are available in every network traversed by the channel. This occurs in the second phase of the setup where each PoP, starting from the destination PoP, sends to the previous-hop PoP a "commit" message. Note that when a PoP determines the set of low priority channels to preempt, it includes this information in the setup request that it sends to the next hop-PoP. When, the latter processes the request, it first "conditionally" preempts these channels and then determines local resource availability. If it determines that it has to preempt some additional low priority channels, it adds them to the list of preempted channels, and propagates the updated list to the next hop-PoP. In this manner, we minimize the number of channels preempted by the PoPs in order to accommodate the new channel.

Note that the signaling protocol we developed supports the setup, release, and QoS modification of point-to-point and point-to-multipoint channel and topology modification (i.e., add/remove sink) of point-to-multipoint MetaNet channels.

3   Resource Management Algorithms

This section presents a short description of two key resource management algorithms that have been implemented in the MetaNet prototype: QoS provisioning algorithm and bandwidth computation algorithm. More details about these algorithms including experiment results on the performance and scalability of the algorithms can be found in [6].

The QoS provisioning algorithm allocates resources to channels taking into account QoS, channel start time, channel end time, and priority. In general, the problem of QoS provisioning is difficult due mainly to the multi-dimensional nature of QoS constraints, i.e., bandwidth, delay, and loss. Multiple constraints often make QoS provisioning intractable; for example, QoS provisioning to accommodate a channel with two independent constraints is NP-complete [7]. MetaNet ads complexity to this difficult problem by allowing the specification of start time, stop time, and priority for channels. To accommodate a channel, the QoS provisioning algorithm should preempt low-priority channels only if it cannot accommodate the channel otherwise. Our goal is to define and implement an algorithm that can be used by a PoP to allocate resources for a requested channel satisfying the channel QoS, time and priority constraints while remaining tractable. This algorithm achieves reduction in computational complexity by decomposing the process of QoS provisioning into multiple steps. First, the algorithm determines the network links that satisfy the bandwidth and loss requirements of the channel over the required time interval. Then, it computes the shortest path in terms of delay. If no such path can be found, the algorithm repeats the same process but considering, this time, the network links that satisfy the bandwidth and loss requirements assuming that all channels with lower priority than the new channel are preempted. With this approach, we transform the NP-complete problem to a problem that can be solved by Dijkstra's algorithm [8]. Such a transformation comes with a penalty: instead of computing a path that optimizes every QoS parameter (e.g., a path with the maximum available bandwidth and the smallest delay that satisfies the requested QoS), we compute a path that optimizes only one QoS parameter and satisfies the other parameters (e.g., a path, if it exists, with the smallest delay that satisfies the requested bandwidth). The most related work on this topic was done by Wang et al.

[9]. However, there are important differences, which distinguish our work from [9]. The problem solved in [9] is limited to the determination of a path that satisfies the bandwidth and delay requirements; it does not consider the connection start time and stop time. More importantly, if no path is found, the algorithms in [9] generate a rejection. On the contrary, our algorithm considers preemption of lower-priority channels to accommodate the new channel.

A PoP uses the QoS provisioning algorithm to check the availability of resources to setup a MetaNet channel within the associated network: It determines a path, if it exists, that satisfies the requested bandwidth and delay. Then, it communicates, via the adapter, with the underlying network to setup the path/connection. The details of this process depends on the type of the underlying network associated with the PoP. In the following, we briefly present these details in the case of ATM network and MPLS-capable IP network [4].

ATM Network:

The PoP mimics the QoS provisioning scheme (i.e., determining a path that satisfies the requested QoS), used by ATM network; this is required because ATM network does not allow route enforcement (SVCs). With this approach, we make sure that the PoP's decisions (accept/reject) match ATM network decisions.

MPLS-Based IP Network:

The PoP can use the QoS provisioning scheme, described above, to compute a path that satisfies the requested QoS. Then, it enforces the setup of this path by setting the corresponding Label switched Path (LSP)/tunnel in the ingress router of the MPLS-capable IP network. The resources reservations and label distributions can be performed using the extended version of RSVP [13].

The second algorithm, *bandwidth computation* algorithm, computes the availability of bandwidth between two hosts that may be in different networks. Two kinds of bandwidth measures are computed: *maximum channel capacity* and *maximum aggregate capacity* (see section 2.2 for definitions) between any two hosts. MetaNet clients may request either static or dynamic (i.e., over a specific time interval) bandwidth measures. Further, the clients may request these measures either without any preemption or with preemption up to a specific priority level. The main problem with the computation of such bandwidth measures is scalability. A simple approach is for one PoP to act as the coordinator. This PoP solicits from every PoP (including self) the static/dynamic state of the latter's associated network (e.g., network topology, maximum capacity of links, available capacity of links). The coordinator then constructs the end-to-end network state by merging the states of all the networks; then it uses Dijkstra's algorithm [8] to compute the maximum channel capacity and Ford-Fulkerson's algorithm [8] to compute the maximum aggregate capacity. However, this approach is neither scalable (with the size of the networks) nor practical (e.g., a network may not be willing to divulge its topology/configuration to other networks because of security policies). Our goal is to define and implement an approach that improves the scalability of this process and that requires exchange of only aggregate network information between PoPs without loss of precision in bandwidth computation. A novel aspect of our approach is that PoPs exchange only the *necessary information* that is required for the specific bandwidth computation requested by the client. No other (either abstract or raw) information about the associated networks is exchanged. For example, to compute the maximum channel capacity, each PoP uses Dijkstra's algorithm to compute the maximum channel capacity between every pair of its border nodes or between the source/destination and every border node (in the case of source/destination PoP). Then, the source PoP constructs an abstract global network view (based on the information produced by the other PoPs) that is represented as a graph where the nodes are either border nodes or hosts. Finally, The source PoP uses Dijkstra's algorithm on the constructed graph to compute the maximum channel capacity between the hosts. A more detailed description of the algorithm can be found in [6].

The most related work to this bandwidth computation problem is the state aggregation scheme used in ATM PNNI [10]. However, there are important differences, which distinguish our work from PNNI. First, in PNNI, the aggregation of network information does not take into account the priority or time schedule of channels/connections. Our approach uses aggregation of network information taking into account priority and time schedule; in fact, our algorithms produce aggregate network information assuming that all channels that have priority lower than a given priority are preempted. Second, in PNNI, each network aggregates and propagates the network state information. On the contrary, our approach requires each network to aggregate and propagate only the necessary information (i.e., a subset of the network state) to support bandwidth computation. Third, in PNNI, the aggregation of network information produces imprecise information. Our approach uses lossless information aggregation.

4   MetaNet Prototype Software Architecture

We have implemented a prototype MetaNet system based on the concepts described in Section 2. Figure 2 illustrates the software architecture of the prototype system. In the prototype architecture, the PoP component described in Section 2 is divided into two parts: one *MetaNet PoP*, and one or more *Network PoPs*. The MetaNet PoP receives client requests. Running under the control of the MetaNet PoP is one or more Network PoPs, one Network PoP for each network under the PoP. For each network under the PoP, there exists an *Adapter* that shields from the PoP technology specific QoS mechanisms and interfaces of the associated network.

The Network PoP is the core component of the MetaNet prototype; it contains four components that interact with each other. The *Internetwork Routing* module computes internetwork routes for MetaNet channels. In the current prototype, routing is topology based; topology changes are exchanged between peer routing modules (see Figure 2). QoS routing is a topic of ongoing research. The Network State Maintainer module maintains the network state needed to support MetaNet functionality. It supports sophisticated state filtering functions.

Thus, for example, it provides an operation for retrieving the state of a network path (sequence of inter-switch links) for a specific time interval considering only channels below a specific priority level. Such state filters simplify the QoS provisioning algorithm used by the Network PoP. The Network Channel QoS Manager module provisions QoS for the channels in the local network. The MetaNet Channel Manager module implements the inter-PoP signaling protocol.

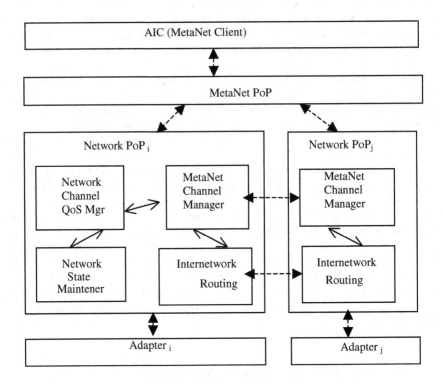

Fig. 2. MetaNet Prototype Software Architecture

Each Adapter provides to the corresponding Network PoP the configuration of the underlying network, including its topology, link capacities, and QoS policies such as bandwidth partitioning among multiple traffic classes. Further, it provides primitives for setting up channels in the local network hiding the technology specific signaling mechanisms and management interfaces.

All interactions between AIC, MetaNet PoP, network PoPs, and adapters occur through CORBA interfaces.

4.1   ATM Adapter

To illustrate how an Adapter hides technology specific details from the PoP, we describe, in this section, the implementation of the ATM network adapter in our prototype.

The ATM Adapter consists of three modules: *Network Mapper*, *Network Channel Scheduler*, and *Channel Policy Server*. See Figure 3. The Network Mapper autodiscovers from the ATM switches (via their SNMP MIBs) the network topology and link capacity information and presents this information to the Network PoP via a CORBA interface.

The adapter views the QoS and scheduling information on channels provisioned by the PoP as policy information. This information is stored in the Channel Policy Server, which is implemented using an LDAP (Light Weight Directory Access Protocol) server [11]. When the Network PoP provisions a channel, it communicates the channel QoS and schedule information to the Adapter using a CORBA interface provided by the Network Channel Scheduler module. This module then stores this information in the policy server using the LDAP API.

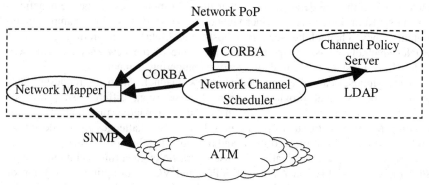

Fig. 3. ATM Network Adapter Architecture

The Network Channel Scheduler periodically accesses the policy server looking for channels that should be started or released. If such channels exist, the scheduler sets up or releases the channels in the network by invoking the Network Mapper via a CORBA interface. The Network Mapper realizes the setup/release of channels by creating/deleting ATM SoftPVCs using SNMP primitives [12]. Note that the Network Channel Scheduler looks at the start time and stop time of a channel. Based on that information, it will either store it in its cache or send it to the policy server for storage and retrieval depending on how far in time it has to perform the setup/release action. The requests in the cache are the ones that will be scheduled for action in the near term.

5   Conclusion

This paper described the functional and software architecture of MetaNet, a network resource management system that provides comprehensive QoS management capabilities for mission critical military applications. It supports current and advance resource reservation for communication channels spanning multiple heterogeneous networks, priority based resource preemption, and queries on static and dynamic end-to-end network resource constraints. MetaNet consists of two major functional components: PoPs and Adapters. Each PoP manages one or more networks, provides MetaNet functions within its domain, and cooperates with other PoPs to support end-to-end services. Adapters hide network specific QoS mechanisms from the PoPs. A prototype based on this architecture has been implemented. See [6] for a description of the resource management algorithms used in this prototype. While this prototype serves as an initial proof-of-concept system, there are some issues that merit further investigation:

Integrated experiments combining AIC and MetaNet are needed to verify whether the MetaNet Link dependency notion is sufficient to support AIC's resource planning functions. If the internetwork topology is not rich enough to support significant route diversity, the dependency information will not be very useful, as each link would then be dependent on other links. In such cases, it may be beneficial for MetaNet to provide to AIC a finer view of the network state. For example, given a potential new channel between two hosts, the MetaNet can return the set of MetaNet channels whose routes intersect the route of the new channel. This will enable AIC to make resource tradeoffs among competing channels.

When MetaNet cannot accommodate a new channel within the requested QoS region, we believe that providing options to AIC (e.g., different start time or different QoS) will enable AIC to make better resource planning decisions. Extending MetaNet with such options introduces new challenges in the design of MetaNet resource allocation algorithms and the QoS signaling protocol.

Currently, we are investigating the use of traffic-engineering techniques, (e.g., MPLS traffic trunks [4] and ATM with VPCs and VCCs) to optimize the setup of MetaNet channels in terms of signaling and traffic multiplexing. For example, instead of setting one VCC (resp. Label Switched Path: LSP) per MetaNet channel, we can setup only one VP (resp. LSP) to transmit the multiplexed traffic of a bundle of MetaNet channels. Note that the MetaNet channels in the bundle are not required to have the same source and destination; VPs that transmit the traffic of the bundle of MetaNet channels will be setup between network devices that are traversed by the bundle of MetaNet channels. The determination of the bundle of the MetaNet channels and the values of the attributes of the VPs (resp. LSPs), such as end points, bandwidth, setup time and priority, depend on many factors including the attributes of MetaNet channels, the state of the network and the aimed trade-off optimization/overhead. Note that with MPLS, MetaNet has more control since it can enforce routes for LSPs (explicit routes are provided to the ingress routers); this is not true with ATM's SVCs.

References

1) DARPA BAA 98-26, AGILE Information Control Environment, June 1998.
2) R. Braden, D. Clark and S. Shenker, Integrated Services in the Internet Architecture: an Overview, RFC 1633, 1994
3) D. Black et al., An Architecture for Differentiated Services, RFC 2475, Dec. 1998
4) E. Rosen, A. Viswanathan, and R. Callon, Multiprotocol Label Switching Architecture, Internet draft <draft-ietf-mpls-arch-06.txt> Aug. 1999
5) ATM User-Network Interface (UNI) Signaling Specification, Version 4, af-sig-0061.000, ATM Forum
6) A. Hafid, N. Natarajan, B. Falchuk, A. Roy, J. Pastor and Y. Kim, Network Resource Management Algorithms for Information Delivery and Planning with Quality of Service Guarantees, Proc. of IEEE MILCOM'00, October 2000

7)   M. Garey and D. Johnson, Computers and Intractability: A Guide to the Theory of NP-Completeness, NY, W.H. Freeman and Co., 1979.

8)   R. Ahuja, T. Magnanti, and J. Orlin. Network Flows: Theory, Algorithms and Applications. Prentice Hall, 1993.

9)   Wang and J. Crowcroft, QoS Routing for Supporting Resource Reservation, IEEE Journal on Selected Areas in Communications, September 1996.

10) ATM Forum, Private Network Network Interface (PNNI), v1.0 specification, May 1996.

11) The OpenLDAP Project, The OpenLDAP Foundation, 1998-1999.

12) ForeRunner ATM Switch Configuration Manual Software version 6.0, 1995-1999, Fore Systems, Inc.

13) B. Davie et al., Use of Label Switching with RSVP, IETF MPLS Working Group, draft-ietf-mpls-rsvp-00.txt, 1998

A Multi-agent System for Resource Management in Wireless Mobile Multimedia Networks

Youssef Iraqi and Raouf Boutaba

University of Waterloo, DECE, Waterloo, Ont. N2L 3G1, Canada
{iraqi,rboutaba}@bbcr.uwaterloo.ca

Abstract. This paper introduces a multi-agent system for resource management developed for cellular mobile networks. The main feature of the proposed multi-agent system is a more efficient support for mobile multimedia users having dynamic bandwidth requirements. This is achieved by reducing the call dropping probability while maintaining a high network resource utilization. A call admission algorithm performed by the multi-agent system is proposed in this paper and involves not only the original agent (at the cell handling the new admission request) but also a cluster of neighboring agents. The neighboring agents provide significant information about their ability to support the new mobile user in the future. This distributed process allows the original agent to make a more clear-sighted admission decision for the new user. Simulations are provided to show the improvements obtained using our multi-agent system.

1 Introduction

Cellular mobile networks have to continue supporting their mobile users after they leave their original cells. This poses a new challenge to resource management algorithms. For instance a call admission process should not only take into consideration the available resources in the original cell but also in neighboring cells as well.

Mobile users are in a growing demand for multimedia applications, and the next generation wireless networks are designed to support such bandwidth greedy applications. The (wireless) bandwidth allocated to a user will not be fixed for the lifetime of the connection as in traditional wireless networks, rather the base station will allocate bandwidth dynamically to users. The Wireless ATM and the UMTS standards have proposed solutions to support such capability.

In this paper we propose a Multi-Agent system for call admission resource management designed for wireless mobile multimedia networks. The call admission process involves not only the cell that receives the call admission request but also a cluster of neighboring cells. The agents share important resource information so the new admitted user will not be dropped due to handoffs. Consequently, the network will provide a low call dropping probability while maintaining a high resource utilization.

The paper is organized as follows. In section 2, we describe the multi-agent

A. Ambler, S.B. Calo, and G. Kar (Eds.): DSOM 2000, LNCS 1960, pp. 218–229, 2000.

architecture proposed in this paper. Section 3 defines the dynamic mobile probabilities used by our multi-agent system. In section 4 we present the call admission
process performed locally by agents in our system. Section 5 introduces the overall admission process involving a cluster of agents. Section 6 gives a description
of agent's cooperation. Section 7 discusses the conducted simulation parameters
and results. Finally, section 8 concludes this paper.

2   The Multi-agent Architecture

We consider a wireless/mobile network with a cellular infrastructure that can
support mobile terminals running applications which demand a wide range of
resources. Users can freely roam the network and experience a large number
of handoffs during a typical connection. We assume that users have a dynamic
bandwidth requirement. The wireless network must provide the requested level
of service even if the user moves to an adjacent cell. A handoff could fail due
to insufficient bandwidth in the new cell, and in such case, the connection is
dropped.

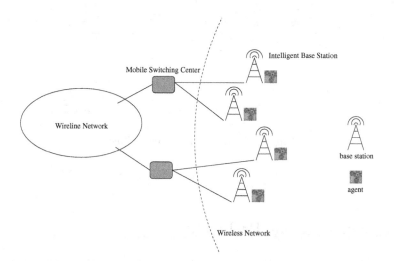

Fig. 1. A Wireless Network and the Multi-agent System

To reduce the call dropping probability, we propose a multi-agent system that
allows neighboring cells to participate in the decision of a new user admission.
Each cell or base station has an agent running on it. The agent keeps track of
the cell's resources and shares information with neighboring agents to better
support mobile users. Each involved agent in an admission request will give its
local decision according to its available resources and information from other

agents and finally the agent at the cell where the request was issued will decide if the new request is accepted or not. By doing so, the new admitted connection will have more chances to survive after experiencing handoffs.

We use the notion of a cluster similar to the shadow cluster concept [5]. The idea is that every connection exerts an influence upon neighboring base stations. As the mobile terminal travels to other cells, the region of influence also moves. The set of cells influenced by a connection are said to constitute a cluster (see figure 2). Each user[1] in the network, with an active connection has a cluster associated to it. The agents in the cluster are chosen by the agent at the cell where the user resides. The number of agents of a user's cluster depend on factors such as user's current call holding time, user's QoS requirements, terminal trajectory and velocity.

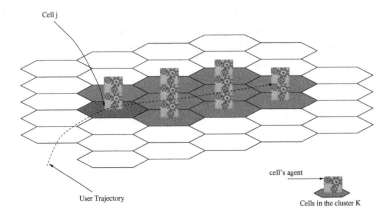

Cell j

cell's agent

User Trajectory

Cells in the cluster K

Fig. 2. Example of a User's Cluster

3   Dynamic Mobile Probabilities

We consider a wireless network where the time is divided in equal intervals at $t = t_1, t_2, ..., t_m$. Let j denote a base station (and the corresponding agent) in the network, and x a mobile terminal with an active wireless connection. Let $K(x)$ denote the set of agents that form the cluster for the active mobile terminal x. We denote $P_{x,j,k}(t) = [P_{x,j,k}(t_0), P_{x,j,k}(t_1), ..., P_{x,j,k}(t_{m_x})]$ the probability that mobile terminal x, currently in cell j, to be active in cell k, and therefore under the control of agent k, at times $t_0, t_1, t_2, ..., t_{m_x}$. $P_{x,j,k}(t)$ represents the projected probabilities that a mobile terminal will remain active in the future and at a particular location. It is referred to as the Dynamic Mobile Probability

---

[1] In the rest of the paper the term "user" and "connection" are used interchangeably

(DMP) in the following. The parameter m_x represents how far in the future the predicted probabilities are computed. It is not fixed for all users and can depend of the user QoS or the actual connection elapsed time.

Those probabilities may be function of several parameters such as: residence time of mobile x in cell j, handoff probability, the distribution of call length for a mobile terminal x when using a given service class, cell size and user mobility profile.

Of course, the more information we have, the more accurate are the probabilities, however the more complex is their computation.

For each user x in the network, the agent that is responsible for, decides the size of the cluster $K(x)$, those are the agents involved in the admission process, and sends the DMPs to all members in $K(x)$. The agent must specify if the user is a new one (in which case the agent is waiting for responses from the members of $K(x)$) or not.

DMPs could range from simple probabilities to complex ones. Simple probabilities can be obtained by assuming, for example, that call length is exponentially distributed, that the call arrival process follows a Poisson distribution and so on.

DMPs can also be complex for example by including information about user mobility profiles. A method for computing dynamic mobile probabilities taking into consideration mobile terminal direction, velocity and statistical mobility data, is presented in [2]. Other schemes to compute these probabilities are presented in [3] [4]. To compute these probabilities, one can also use mobiles' path/direction information readily available from certain applications, such as the route guidance system of the Intelligent Transportation Systems with the Global Positioning System (GPS).

4   Local Call Admission Decision

User's traffic can be either voice, data or video. Voice users are usually characterized by a fixed bandwidth demand. Data and video users have a dynamic bandwidth requirement due to the burstiness of the carried traffic. Without loss of generality, we assume that all users are characterized by a bandwidth demand distribution $f_x(E_x(c), \sigma_c)$. Where $E_x(c)$ and σ_c are the mean and the standard deviation of the distribution f_x respectively, and c is user's x type of traffic. $E_x(c)$ depends of user x traffic type c (voice, data or video).

In conjunction with the emergence of adaptive multimedia encoding [6] [7] [8], QoS adaptation schemes have been proposed to reduce handoff drops. In these schemes a connection's QoS can be downgraded if the available bandwidth in the new cell is not sufficient [9] [4]. Such schemes can be easily integrated in our system as part of the local call admission decision.

4.1   Computing Elementary Responses

At each time t_0 each agent, in a cluster $K(x)$ involved in our call admission (CA) process for user x, makes a local CA decision for different times in the future

$(t_0, t_1, ..., t_{m_x})$. Based on these CA decisions, we call Elementary Responses, the agent makes a final decision which represents its local response to the admission of user x in the network. Elementary responses are time dependent. The computation of these responses is different according to the user location and type. The user can be either a local new user or a new user that has a non null probability to be in this cell in the near future.

User Types. An agent may be involved in the processing of different types of user. Possible user types at time t_0 are:

1. Old users local to the cell
2. Old users coming from another cell (executing a handoff)
3. New users (at time t_0) from within the cell
4. New users (at time t_0) from other cells

New users are defined as all users seeking admission at time t_0. Users of type 1 have the highest priority. Priority between other users is subject to some ordering policy. The network try to support old users if possible and uses the DMPs to check if a cell can accommodate a new user who will possibly come to the cell in the future.

Local Call Admission Decision at Time t_0 for Time t_0. An agent can apply any local call admission algorithm to compute the elementary responses. In this work we assume that the agents use the Equivalent Bandwidth approach to compute these responses. Example of such a scheme is described in [1]. Other schemes can be downloaded to the agents from the management station.

The processing of local new users will be explained in section 5.

Local Call Admission Decision at Time t_0 for Time t_l $(t_l > t_0)$. Each agent computes the equivalent bandwidth at different times in the future according to the DMPs of future users.

If user x, in cell j at time t_0, has a probability $P_{x,j,k}(t_l)$ to be active in cell k at time t_l and has a bandwidth demand distribution function $f_x(E_x(c), \sigma_c)$, then agent k should consider a user x', for time t_l, with a bandwidth demand distribution function $f_{x'}(E_x(c) \times P_{x,j,k}(t_l), \sigma_c)$ and use it to make its local call admission decision.

We denote $r_k(x, t)$ the elementary response of agent k for user x for time t. The agent sets in which order of users it will perform its call admission process. For instance, the agent can sort users in a decreasing order of their DMPs. If we assume that user x_i has higher priority than user x_j for all $i < j$, then to compute elementary responses for user x_j, we assume that all users x_i with $i < j$ that have a positive elementary response are accepted. As an example, if an agent wants to compute the elementary response r for user x_4, and we have already computed r for users $x_1 = 1$, $x_2 = 1$ and $x_3 = 0$, then to compute r for x_4 the agent assumes that user 1 and 2 are accepted in the system but not user x_3.

We propose also that the agent reserves some bandwidth in case of an erroneous prediction. This amount of reserved bandwidth is a parameter of our scheme and can be tuned to have the best performance. The choice of this parameter depends on the precision of the DMPs.

4.2 Computing the Final Responses and Sending the Results

Since the elementary responses for future foreign users are computed according to local information about the future, they should not be assigned the same confidence degree. Indeed, responses corresponding to the near future are more likely to be more accurate than those of the far future.

We denote $C_k(x,t)$ the confidence that agent k has about its elementary response $r_k(x,t)$. The question arises on how the agent can compute (or simply choose) the confidence degrees $C_k(x,t)$, typically between 0% and 100%. One way to compute the confidence degrees is to use the percentage of available bandwidth when computing the elementary response as an indication of the confidence the agent may have in this elementary response.

If for user x, agent k has a response $r_k(x,t)$ for each t from t_0 to t_m with a corresponding DMPs $P_{x,j,k}(t_0)$ to $P_{x,j,k}(t_m)$, then to compute the final response those elementary responses are weighted with the corresponding DMPs. The final response from agent k to agent j concerning user x is then :

$$R_k(x) = \frac{\sum_{t=t_0}^{t=t_{mx}} r_k(x,t) \times P_{x,j,k}(t) \times C_k(x,t)}{\sum_{t=t_0}^{t=t_{mx}} P_{x,j,k}(t)} \tag{1}$$

where $C_k(x,t)$ is the confidence that agent k has about the elementary response $r_k(x,t)$. To normalize the final response each elementary response is also divided by the sum over time t of the DMPs in cell k. Of course, the sum $\sum_{t=t_0}^{t=t_{mx}} P_{x,j,k}(t)$ should not be null (which otherwise means that all the DMPs for cell k are null!). Agent k, then, sends the response $R_k(x)$ to the corresponding agent j.

5 Taking the Final Decision

Here the decision takes into consideration the responses from all agents in the user's cluster. The admission process concerns only new users seeking admission to the network and not already accepted users.

We assume that agent j has already decided the cluster $K(x)$ and that agent j has already assigned to each agent k in the cluster $K(x)$ a weight $W_k(x)$. Each weight represents the importance of the contribution of the associated agent to the global decision process. Usually an agent that is involved more in supporting the user has a high weight value. Weights $W_k(x)$ depend on the DMPs and the time t.

We suggest to use the following formula to compute the weights $W_k(x)$:

$$W_k(x) = \frac{\sum_{t=t_0}^{t=t_{mx}} P_{x,j,k}(t)}{\sum_{k' \in K} \sum_{t=t_0}^{t=t_{mx}} P_{x,j,k'}(t)} \tag{2}$$

If we assume that each response $R_k(x)$, from agent k, is a percentage between 0% (can not be supported at all) and 100% (can be supported), then the agent computes the sum of $R_k(x) \times W_k(x)$ over k.

The final decision of the call admission process for user x is based on

$$D(x) = \sum_{k \in K} R_k(x) \times W_k(x) \tag{3}$$

If $D(x)$ is higher than a certain threshold then, user x is accepted; otherwise the user is rejected. The threshold can be specified by the user. The more higher is the threshold the more likely the user connection will survive in the event of a handoff.

Combining eq. 1 and eq. 2, eq. 3 can be written as:

$$D(x) = \frac{1}{\alpha} \sum_{k \in K} \sum_{t=t_0}^{t=t_{mx}} r_k(x,t) \times P_{x,j,k}(t) \times C_k(x,t) \tag{4}$$

With $\alpha = \sum_{k' \in K} \sum_{t=t_0}^{t=t_{mx}} P_{x,j,k'}(t)$. Only the value $\sum_{t=t_0}^{t=t_{mx}} r_k(x,t) \times P_{x,j,k}(t) \times C_k(x,t)$ should be computed locally in each cell, and the final result is then, simply the sum of all responses from all the agents in the cluster K divided by α.

6   Agent's Cooperation

Each time t, an agent j should decide if it can support new users. It decides locally if it can support users of type 1 and 2 that have higher priority than other type of users (cf. user types in section 4.1). This is because, from a user point of view, receiving a busy signal is more bearable than having a forced termination. The agent also sends the DMPs to other agents and informs them about its users of type 3 (step 2 in figures 3, 4). Only those who can be supported locally are included, other users of type 3 that can not be accommodated locally are rejected. At the same time, the agent receives DMPs from other agents and is informed about users of type 4.

Using equation 1, the agent decides if it can support users of type 4 in the future and it sends the responses to the corresponding agents (step 3 in figures 3, 4). When it receives responses from the other agents concerning its users of type 3, it performs one of the two following steps (step 4 in figures 3, 4): If the agent can not accommodate the call, the call is rejected. If the agent can accommodate the call, then the call admission decision depends on equation 4.

Figure 3 shows the different steps of agent's cooperation when processing an admission request. Figure 4 depicts the admission process diagram at the agent receiving the admission request and at an agent belonging to the cluster. Because the admission request is time sensitive the agent waiting for responses from the agents in the cluster will wait until a predefined timer has expire then he will assume a negative response from all agents that could not respond in time.

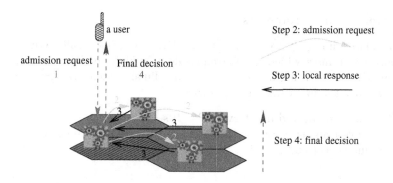

Fig. 3. Agent's Cooperation for the Admission of a User

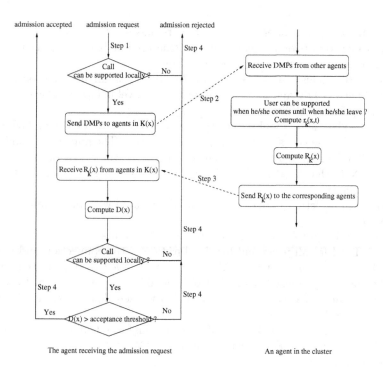

Fig. 4. Admission Process Diagram

7   Performance Evaluation

7.1   Simulation Parameters

For the sake of simplicity, we evaluate the performance of our Multi-Agent system for mobile terminals which are traveling along a highway. This is a simplest environment representing a one-dimensional cellular system. In our simulation study we have the following assumptions:

1. The time is quantized in intervals $T = 10s$
2. The whole cellular system is composed of 10 linearly-arranged cells, laid at 1-km intervals (see figure 5).

Fig. 5. A Highway Covered by 10 Cells

3. During each time interval, connection requests are generated in each cell according to Poisson process. A newly generated mobile terminal can appear anywhere in the cell with equal probability.
4. Mobile terminals can have speeds of: 70, 90, or 105 km/h. The probability of each speed is 1/3, and mobile terminals can travel in either of two directions with equal probability.
5. We consider three possible types of traffic: voice, data, or video. The probabilities of these types are 0.7, 0.2, 0.1 respectively. The number of bandwidth units (BUs) required by each connection type is: voice $= 1$, data $= 5$, video $= 10$. Note that fixed bandwidth amounts are allocated to users for the sake of simplicity.
6. Connection lifetimes are exponentially-distributed with mean value equal to 180 seconds.
7. Each cell has a fixed capacity of 40 bandwidth units.
8. m_x is fixed for all users and for the duration of the connection and is equal to 18. This means that the DMPs are computed for 18 steps in the future.
9. The size of the cluster $K(x)$ is fixed for all users and is equal to 5. This means that four cells in the direction of the user along with the cell where the user resides form the cluster.
10. We simulate a total of 4 hours of real-time highway traffic, with a constant cell load equal to 360 new calls/h/cell.
11. The DMPs are computed as in [2].
12. All users have the same threshold.
13. The confidence degree is computed as follows: $Confidence = e^{(1-p)} * p^3$ where p is a real number between 0 and 1 representing the percentage of available bandwidth at the time of computing the elementary response.

7.2 Simulation Results

In our simulations, a user x requesting a new connection is accepted into a cell only if the final decision $D(x)$ is above an acceptance threshold value. We varied this threshold value to observe its effect on the call dropping percentage and the average bandwidth utilization in the cells of the network.

By varying the value of the threshold in the simulations, we were able to decrease the percentage of dropped calls while maintaining a good average bandwidth utilization.

Figure 6 depicts the average bandwidth utilization of the cells in the network, and the corresponding percentage of dropped calls for different acceptance threshold values.

The top curve represents the average number of BU's that are used in all

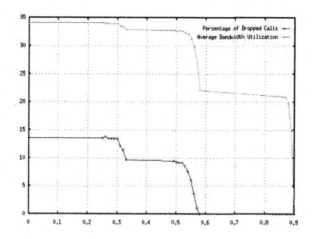

Fig. 6. Average Bandwidth Utilization and Percentage of Dropped Calls According to the Acceptance Threshold Value

cells in the network, considering the entire simulation time. When the threshold is equal to zero, the average bandwidth utilization is at its maximum value. In this case, the maximum bandwidth utilization is approximately equal to 34 BU's. The bottom curve depicts the percentage of dropped calls in the network. The highest percentage of dropped calls also occurs when the threshold is equal to zero; in this case, all connection requests are accepted regardless of the final decision $D(x)$, as long as there is available bandwidth in the cells where the connections are requested. For the simulated cell load, the maximum percentage of dropped calls is equal to 14%. By adjusting the threshold value, our Mutli-Agent system can control the percentage of calls that will be dropped. For example, with a threshold value of 57%, the percentage of dropped calls is reduced to the value of 1% while maintaining at the same time a high average bandwidth utilization value of 27 BUs. The proposed scheme allow a tradeoff between average bandwidth utilization and the percentage of dropped calls. If the threshold value is 83% then no calls need to be dropped with a corresponding average bandwidth utilization of 21 BUs. Thus, the proposed scheme can reduce the percentage of dropped calls with an acceptable degradation in total bandwidth utilization.

8   Conclusion

In this paper, we have described a Multi-Agent system for resource management suitable for wireless multimedia networks. The proposed system operates in a distributed fashion by involving, in a call admission decision, not only the agent receiving the admission request, but also a determined number of neighboring agents. The goals underlying the design of our algorithm are: (1) to support mobile multimedia users with dynamic bandwidth requirements; (2) to reduce the call dropping probability while maintaining a high network resource utilization; and (3) to distribute call admission decision among clusters of neighboring agents to allow more clear-sighted decisions and hence a better user survivability in the network. More technically, our algorithm can integrate easily any method for computing Dynamic Mobile Probabilities (DMPs). It can also rely on different local call admission schemes including those designed for adaptive multimedia applications. Those schemes can be downloaded to the agents by the management system.

Simulations results have shown that by implementing the proposed multi-agent system, the wireless netwok is able to lower the call dropping probability while offering a high average bandwidth utilization. The wireless network is also able to maintain a high acceptance probability for new users. The signaling load induced by agent's communication is considered here acceptable as far as it only involves few messages exchanged between agents through the wired network which is assumed to be of high capacity. More simulations are envisaged in the future to evaluate our multi-agent system in more sophisticated situations, for example with users having dynamic bandwidth requirements, cell loads, and traffic distributions. Also envisaged is studying the influence of the number of agents involved in a call admission decision.

References

1. J. Evans and D. Everitt, 'Effective bandwidth based admission control for multi-service CDMA cellular networks,' IEEE Trans. Vehicular Tech., Vol 48, No 1, pp 36-46, January 1999.
2. D. A. Levine, I. F. Akyildz and M. Naghshineh, 'A Resource Estimation and Call Admission Algorithm for Wireless Multimedia Networks Using the Shadow Cluster Concept,' IEEE/ACM Transactions on Networking, vol. 5, no. 1, Feb. 1997.
3. Sunghyun Choi and Kang G. Shin, 'Predictive and Adaptive Bandwidth Reservation for Hand-Offs in QoS-Sensitive Cellular Networks,' in Proc. ACM SIG-COMM'98, pp. 155-166, Vancouver, British Columbia, September 2-4, 1998.

4. Songwu Lu and Vaduvur Bharghavan, 'Adaptive Resource Management Algorithms for Indoor Mobile Computing Environments,' in Proc. ACM SIGCOMM'96, pp. 231-242, August 1996.
5. D. A. Levine, I. F. Akyildz and M. Naghshineh, 'The shadow cluster concept for resource allocation and call admission in ATM-based wireless networks,' in Proc. ACM Int. Conf. Mobile Comp. Networking MOBICOM'95, Berkeley, CA, pp. 142-150, Nov. 1995.
6. R. Rejaie, M. Handley, and D. Estrin, 'Quality Adaptation for Congestion Controlled Video Playback over the Internet,' in Proc. ACM SIGCOMM'99, September 1999.
7. S. McCanne, M. Vetterli, and V. Jacobson, 'Low-Complexity Video Coding for Receiver-Driven Layerd Multicast,' IEEE Journal on Selected Areas in Communications, Vol. 15, No. 6, pp. 983-1001, August 1997.
8. J. Hartung, A. Jacquin, J. Pawlyk, and K. Shipley, 'A Real-time Scalable Video Codec for Collaborative Applications over Packet Networks,' ACM Multimedia'98, pp. 419-426, Bristol, September 1998.
9. K. Lee, 'Supporting mobile multimedia in integrated service networks,' ACM Wireless Networks, vol. 2, pp. 205-217, 1996

Using Message Reflection in a Management Architecture for CORBA

Maarten Wegdam[1,3], Dirk-Jaap Plas[1], Aart van Halteren[2,3], and Bart Nieuwenhuis[2,3]

[1] Lucent Technologies - Bell Labs Twente, Capitool 5, 7521 PL Enschede, The Netherlands
{wegdam,dplas}@lucent.com
[2] KPN Research, PO Box 96, 7500 AB Enschede, The Netherlands
{A.T.vanHalteren,L.J.M.Nieuwenhuis}@kpn.com
[3] Faculty of Computer Science, University of Twente, Enschede, The Netherlands

Abstract. The availability of object middleware, such as CORBA, is rapidly being accepted as a means for cost effective and fast development for a wide range of distributed applications. Distributed applications that are built using these technologies often comprise many objects and become more and more complex. The deployment of such large distributed applications requires a significant improvement of management methods and tools. In this paper, we present a management architecture for object middleware based systems. We use message reflection to extend the middleware layer with management capabilities, i.e. we monitor the application by observing the messages exchanged between the objects of the distributed application. We argue why management should be transparent to the application developer and show that message reflection supports this management transparency. We have compared different mechanisms to implement message reflection in CORBA, and argue why portable interceptors are the most suitable. Finally, we describe our prototype and the lessons we learned.

1. Introduction

In order to keep a deployed system in an operational and usable state, capabilities for the different areas of management have to be provided. These areas are fault management, configuration management, accounting management, performance management, and security management [9]. Object middleware technology does not offer sufficient management capabilities in any of these areas.

We believe that management should be dealt with in a generic manner, and thus should not be solved by the application developer. The application developer should only be concerned with the functional behavior of the software under development, and not with the management issues. The benefits of solving the management problem in a generic manner are that it only has to be solved once in stead of again and again for every new component or application. This reuse of code reduces, among others, development costs and time-to-market. Ideally, when developing and

A. Ambler, S.B.Calo, and G. Kar (Eds.): DSOM 2000, LNCS 1960, pp. 230 - 242, 2000.

deploying a new application the required management functionality is inherently present.

In this paper, we describe how to add management to an object middleware system in a transparent manner, effectively extending the distribution transparencies with a new management transparency. We propose to do this by using message reflection, i.e. intercepting and reflecting on the messages that are sent between the different components of an application.

In Section 2, we describe what we mean with management of middleware, the different roles we distinguish, and what the different roles require from the management system. Section 3 describes related work in this area. Section 4 compares different mechanisms to implement message reflection in CORBA. Section 5 describes our management architecture, how it uses message reflection, and our prototype. Section 6 describes the lessons learned and Section 7 finally describes our conclusions and future work.

2. Management of CORBA

We divide the management of object middleware in two separate but related issues:

- Management of the ORB itself.
  Examples are the number of threads that are available, how many threads are actually used, what network resources are available, queue sizes, the number of registered objects, used policies and the lifecycle of object adaptors.
- Management of the objects running on top of it.
  This is partly object specific and partly generic. We only consider the generic aspects in this paper. Examples are the availability of an object, i.e. whether an object is alive or not and able to respond to requests, in what stage of its lifecycle it is, what the delay on requests is, detection of user- or system exceptions that occur, the logging of requests, and the uptime.

The requirements for a management architecture for object middleware, like CORBA, can be derived by considering the parties involved in developing and deploying object middleware applications. We distinguish four different roles:

- The application or component developer
- The system administrator
- The management tool vendor
- The ORB vendor

Fig. 1 depicts who develops which part of the object middleware and management system. A different texture indicates a different role. Please note that the instrumentation is collocated with the managed entities.

The *component developer* focuses on the business logic of the component, and should be masked as much as possible from the technology specific details of the object middleware platform. The main purpose of object middleware is to provide distribution transparencies [23] such as location and access transparency. From the application developer's perspective, management of the application is not part of the business logic, and should be dealt with by the middleware. Application management should be transparent to the application developer. We refer to this as management transparency.

A *system administrator* of a
multi-vendor middleware
environment could be faced
with different, and possibly
incompatible, ways that ORB
and application vendors have
made their software
manageable. This is not
acceptable, because the system
administrator can not be
expected to learn all the
vendor specific features in
order to be able to manage a
middleware-based application.
Therefore, both the ORB and
the components running on top

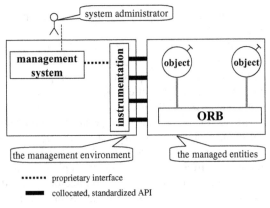

Fig. 1. The Different Roles

of it must be manageable in a uniform manner. No matter who the ORB or component
vendor is, the management view should be equal. A second requirement is that the
management should demand a minimal effort from the system administrator, both
when installing the management tools, and at run-time. In addition, a system
administrator may want to create an integrated view covering network management,
application management, and middleware management.

A *vendor of management tools* wants to provide his management solution as a third
party add-on to ORB implementations of different vendors without adaptation or
negotiation of a special management interface to a specific ORB. This is comparable
to the requirement of a system administrator for uniform management. It requires a
standardized (management) interface to an ORB to add management functionalities.

ORB vendors may not want to get into the business of ORB management tools but
still want to provide a manageable product and allow the customer to choose their
favorite management tool. This also requires a standardized (management) interface
on the ORB.

3. Related Work

The management of networks and network elements has been an active research area
for a relatively long time, resulting in mature products and standards. Examples of
these are the Telecommunications Management Network (TMN)[25], the Common
Management Information Protocol (CMIP)[6] and the Simple Network Management
Protocol (SNMP)[24]. The management of distributed applications recently is getting
more attention from academia and industry, for example within the Distributed
Management Task Force (DMTF)[31]. The management of ORBs and the
components running on top of it are a new area of active research. We mention the
most important papers, standards and projects in both industry and academia.

In the Fachhochschule Wiesbaden (Germany) work was done on management of
distributed systems [15], specifically the monitoring of Orbix based applications with
the so-called ObjectMonitor. They use Orbix proprietary filters to intercept in and

outgoing requests, and have an SNMP interface to ObjectMonitor. Current research seems to mostly focus on the automation of management tasks.

A paper from the University of Aachen (Germany) describes the usage of non-co-located proxy servers to intercept in- and outgoing requests [17]. The major benefits of this approach are that it is ORB independent, and does not require recompiling an existing server. However this approach is relatively inflexible, it has to be done manually, and it probably introduces substantial overhead. Other papers, especially [16], focus more on the use of agents for management of CORBA, without explicit consideration on how to instrument the ORB and without providing a design.

In [28] three CORBA management tools for Orbix are compared. These are OrbixManager, CORBA Assistant and Object/Observer. They conclude that these three tools focus on fault and configuration management, and that manageable units are commonly CORBA processes. All three tools do instrumentation by using Orbix filters, which requires adding a few lines of application code to activate them.

The best-known management application for CORBA is probably OrbixManager [20]. OrbixManager is a combination of instrumentation to the Orbix ORB and a management service and console. OrbixManager manages the Orbix-based middleware components of an application. It extends the applications with management functionality by linking them with a management library. Some other ORB vendors have also implemented some proprietary management extensions to their ORBs. For example Inprise's Visigenic ORB [32] and BEA's Weblogic Enterprise ORB [33].

Sun, together with some other companies, made a specification called the Java Management Extensions (JMX)[13] for management of Java based application using Java technology. Main features are a push distribution mechanism, usage of JavaBeans, and remote access through several protocols, including SNMP, RMI and IIOP. JMX is based on a product of Sun called the Java Dynamic Management Kit (JDMK)[12]. A claimed benefit of JMX based products, compared to more static solutions, is that the management intelligence can be easily distributed, and can be located with the managed entity. This can reduce network traffic and increase flexibility.

Marvel [2] is a management environment that is comparable with JMX in that it also is Java-based and allows the uploading of management code to agents, and allows the usage of different management protocols. In Marvel however one can define automatically computed views of management information, which is not possible in JDMK. This can increase the scalability of the management systems. Marvel also includes functionality to visualize the management information.

Fosså and Sloman describe in [10] a management system that can be used for configuration management of a CORBA system. The Darwin configuration language is used to describe the initial configuration. A system administrator can change the configuration of the distributed system by altering the binding that exists between the CORBA objects. The objects have to be altered to allow this third-party binding. The paper focuses on the configuration description, the configuration evolution and the GUI. The implications on the object specific code is not very clear, including if this can be done in a CORBA compliant manner

MIMO, MIddleware MOnitor, is an on-line monitoring tools for distributed object-environments [22]. The distributed environment is separated in different layers, each layer is monitored, and information from the different layers is mapped to each other. This approach has not yet been implemented, and the issue of how to instrument the monitored objects is for further research.

Research done at the University of Lancaster [3] uses reflection in middleware. They argue that current ORBs have a pre-defined and mostly standardized behavior, and that reflection can be used to easily construct a customized ORB from more or less independent components to get the behavior that is desired for a specific domain or application. An architecture for reflective middleware is described in which the meta-space of an object is divided in three different meta-models; the compositional meta-model, the encapsulation meta-model and the environment meta-model. A description of an initial implementation of this architecture, implemented in Python, can be found in [8].

4. Message Reflection

One of the major requirements stated in Section 2 is to make the management transparent to the application developer. This means among others that we cannot intertwine management functionality with the core functionality of the object. We propose to exploit the fact that distributed objects interact by exchanging messages. By intercepting and inspecting messages, we can deduct information relevant for the management of these objects. This is also known as message reflection.

In this section, we present the mechanisms for implementing message interception for management of CORBA, and we discuss the relative advantages and disadvantages of each mechanism.

Sniffing

A very straightforward method for intercepting messages is network sniffing. This is typically done by intercepting TCP/IP messages. After filtering out non-relevant messages, the IIOP messages are parsed to determine the GIOP message type and parameters, effectively de-marshalling the requests. The obvious advantage of this method is that it is completely non-intrusive and transparent for the client, the server and the ORB. Disadvantages are that only messages actually passing through the network segment will be sniffed, excluding messages sent between clients and servers on the same host. A second problem is that this method is only practical on a network that uses broadcast technology, such as Ethernet. It would otherwise require a sniffer for each host. A third limitation of this method is that does not allow message to be altered.

Instrumented Stubs and Skeletons

In the normal case of static invocations and interfaces all messages will pass through the stubs and skeletons, which the IDL compiler has generated. Since the stubs and skeletons are always available in source, they can be instrumented to read or even change messages that pass through them. The main disadvantage of this method is that it is very ORB and IDL compiler dependent. Another disadvantage is that messages for dynamic invocations (Dynamic Invocation Interface in CORBA) and dynamic interfaces (Dynamic Skeleton Interface in CORBA) are not intercepted, since dynamic messages do not pass through the stubs or skeletons.

Wrapping

Wrapping is of course a well-known pattern to add functionality to an (existing) object or class. Wrapping can be used to intercept messages going to and from objects in a distributed application. The main advantage of this method is that it is usually transparent to the server object. The problem is that the client has to send requests to the wrapper object instead of the actual object, which is especially difficult when object references are passed between clients. This problem requires a lot of administration and thus introduces a management problem of itself. Also, it introduces a substantial delay. But in a system with a fixed number of objects on fixed locations this can be a solution worth considering.

Inheritance and Delegation

At first glance, it might seem like a good idea to use inheritance to add management intercepting capabilities to an object. One can introduce a new class at the top of the inheritance tree that all other objects inherit from, or one can do the opposite and create a subclass of an object to do the intercepting. The first approach is not suitable for intercepting messages without requiring major changes to the ORB, since the instrumentation will not be in the invocation path. It can be used to intercept lifecycle events on an object. The second approach could be a solution, but introduces so-called inheritance anomalies [1]. It is also quite intrusive to the application object and requires the usage of an object-oriented implementation language. Delegation has similar disadvantages as inheritance, especially since it is intrusive to the application object.

Composition Filters

Composition filters [1] is a modeling concept in which the actual object has explicit incoming and outgoing filters that can manipulate messages, e.g. to delay or to dispatch messages. It allows for a very clean separation of concerns, and solves the problem of inheritance anomalies. Composition filters require support by the implementation language, or even better support by CORBA (for example as an extension to IDL). Unfortunately there is only limited support for this for most implementation languages, and there is certainly no support for it in CORBA/IDL.

CORBA Portable Interceptors

Interceptors were first introduced in CORBA in version 2.2 of the CORBA specification [7]. This specification defines interceptors, which can intercept requests at defined points inside the ORB. This interceptor specification is rather ambiguous, and is about to be superseded by the *portable interceptor* specification. The portable interceptor specification [21] defines two kinds of interceptors: request and IOR interceptors.

Request interceptors are located in the invocation path of all ORB mediated requests, thus also invocations to co-located objects. They can intercept in- and outgoing requests on both the client and the server-side, resulting in a total of four interception points, see Fig. 2.

A request interceptor can affect the outcome of a request by raising a system exception at any of the interception points, or directing a request to a different location. The target and parameters of a request can be inspected, but not altered. Several interceptor instances can be registered for one interception point, in which case they run in sequence. A request interceptor can inspect and alter the ServiceContext information.

IOR interceptors are purely server side, and are called when the ORB is creating an IOR, or to be more precise when it is assembling the list of components that will be included in the IOR. This does not necessarily mean that this interceptor is called for every individual object reference. For example, the Portable Object Adapter (POA) specifies policies at POA granularity and therefore this operation might be called once per POA rather than once per object.

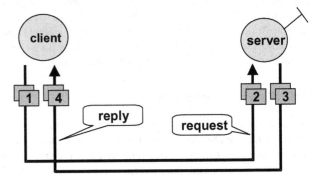

Fig. 2. Request Interceptors and the Invocation Path

At the time of writing some ORBs, like Orbacus [18], have already implemented the portable interceptors. Most other ORBs have either the 'old' CORBA 2.2 request interceptors, or have a similar mechanism, e.g. Orbix' filters [14] or VisiGenic's interceptors [32].

Interceptors in CORBA are relatively non-intrusive, and can be developed by the provider of a CORBA management system and simply be 'plugged in' into any ORB that needs to be managed. It can intercept all the requests going into and out of a CORBA object. The disadvantages are that depending on the programming language it can requires recompilation and a small code change in the application code to activate the interceptors.

Operating System Interceptors

A final mechanism we describe is what we call Operating System (OS) interceptors. These interceptors are positioned between the ORB and OS-level interface to the network. Instead of intercepting a message within the ORB, the messages are intercepted after they leave the ORB, but just before they enter the TCP/IP library. This approach is used in Eternal [19]. The major benefit of this approach is that it is completely transparent to the component programmer and to the ORB implementation. There are however several disadvantages. A major one is that since the intercepted messages are IIOP messages, the information at this level is quite dense and requires reverse processing (i.e. de-marshalling) to obtain request information. Besides this, the method depends on the usage of dynamically linked libraries, and is dependent on the OS and network. Last but not least, requests between co-located objects cannot be intercepted, since they usually bypass the TCP/IP library.

Summary

Although the possibilities for reflection in general within CORBA are limited [27], there are several ways for extending an ORB if we limit ourselves to message reflection. We consider CORBA interceptors to be the most suitable mechanism for this, because they can intercept messages for co-located and remote invocations, without depending on the network or OS. The new portable interceptors provide for an intercepting mechanism that can be used for monitoring, but it has only limited capabilities for control functionality.

5. Architecture

In this section we describe our management architecture. Our management architecture is based on the Manager-Agent paradigm [11]. We compare how this paradigm is used in different management systems. We use this as input for our management architecture. After this we describe and evaluate our management architecture, and our initial implementation.

Manager-Agent Paradigm

The general usage of the manager-agent paradigm in management systems like SNMP and CMIP already has proven its applicability for management. This paradigm is used in two ways.

The first way is to centralize all management functionality with the manager. This approach is followed by the traditional management systems like SNMP and CMIP, and by OrbixManager. The centralization of management functionality however introduces problems with respect to scalability, information overload at the manager and network delays.

The second way is to distribute the management functionality over the manager and the agent. JMX, Marvel and the University of Aachen [16] follow this approach. By distributing the management functionality it is tried to solve the above mentioned problems.

The Manager-Agent paradigm can be implemented in several ways. The major design choices when implementing the Manager-Agent paradigm are [4]:
1) Which technology is used for communication?
2) Which part initiates the data transfer?
3) How are the parts bundled?

Based on these choices two commonly used approaches can be distinguished; the library based agent and the application based agent.

In the library based agent approach the agent and the instrumentation are implemented as a library that is linked to the managed entity. The manager is the only part of the management system that runs outside the library. OrbixManager uses this approach. In the application-based approach the agent is a separate entity, and only the instrumentation is co-located with the managed entity. Both approaches are depicted in Fig. 3.

As explained in [4] the application based approach allows for more of the management functionalities to be implemented in the agent, and is thus more scalable. We are therefore using the application-based approach.

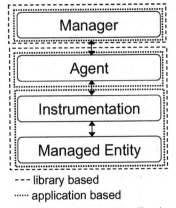

--- library based
····· application based

Fig. 3. Library versus Application Based Approach

The Management Architecture

Based on the requirements mentioned in Section 2 and the above-mentioned issues, we have developed a management architecture as depicted in Fig. 4. In the following subsections we describe each part of the architecture.

The Manager. A management console implements the manager. It provides a Graphical User Interface to the administrator for the whole management system. It provides views for individual managed entities via the IDL interface the managed object offers, and views for groups of management entities.

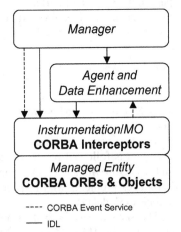

---- CORBA Event Service
—— IDL

Fig. 4. The Management aArchitecture

The Agent. The main responsibility of the agent is to store and enhance management information. The agent provides an IDL interface for the manager to access this information. The managed objects push relevant management events though a

CORBA Event Service to the agent. The agent also uses synchronous communication (IDL) to request specific management information from the managed objects.

The Managed Object / Instrumentation. The managed object provides the agent with a management view on the managed entities. It is implemented using CORBA interceptors and standardized CORBA interfaces.

We use interceptors to monitor in and outgoing requests. The agent uses this data to derive, for example, response times, network failures and client-server relationships.

The ORB internal interfaces are not specified with the intention to be used for management. We do however use the existing POA, Object and ORB interfaces because they do offer some useful management functionalities [26] not available through interceptors.

The Managed Entity. The managed entities are the CORBA ORB core, and the CORBA objects. By managing these, we also manage the applications and services the objects are a part of.

6. Lessons Learned from the Prototype Implementation

We have prototyped our management architecture, and have successfully tested it.

With CORBA interceptors it is possible to develop management functionality independent of the ORB implementation, and thus use our management system for every ORB that implements portable interceptors. We can derive configuration, accounting, performance, and availability information about the application objects. By combining different information items the information processor is able to enhance this information, for example to determine how an application is spread over different hosts.

The control possibilities are limited by the possibilities of the instrumentation, thus the ORB interceptors and ORB internal interfaces. As a consequence our implementation has only limited possibilities for control.

The usage of portable interceptor, and thus the instrumentation, is completely transparent to the component developer because the Java language mapping allows portable interceptors to be added to already compiled code.

The management system can be distributed and, although not discussed in this paper, allows a hierarchical structure. This enhances scalability. The hierarchical structure also provides a way for management of domains of managed entities. We currently use the CORBA Event Service for asynchronous communication due to the lack of a suitable implementation of the CORBA Notification Service. The Notification Service however has a better filter and subscription mechanism, which also will reduce network traffic.

We use a generic event format defined in XML for exchanging management information. This flexible design enables easy integration with existing, e.g. SNMP or CMIP based, or new management systems.

The usage of request level interceptors introduces a significant overhead. Preliminary testing revealed that with JacORB [30] the typical delay overhead is 300 ms per request. This can be minimized by disabling interceptors that produce irrelevant management information.

7. Conclusions

We have described how message reflection can be used to manage object middleware based applications. We have compared different message reflection mechanisms in CORBA, and have selected CORBA interceptors as most suitable. CORBA interceptors can monitor object interactions in a non-intrusive manner, are ORB independent and are transparent to the object developer. We have an initial implementation of a management architecture for the management of CORBA based applications. Besides interceptors, our implementation also uses standardized ORB interfaces for instrumentation.

Based on our experiences we have identified a number of issues for further study. One of the major issues is which resources we should manage and which not. We will evaluate our current choice in different projects that use CORBA, and with experiences gained from these projects we will adjust our current choices.

We did not address the issue of policy based management in this paper. We do believe however that this is the way to go, and are working on using this within our management architecture.

Our architecture assumes one logically centralized manager that controls all the distributed components. For cross-organizational applications this will not be the case, several managers will exist, each independently managing their own domain. This will have consequences for our management architecture.

We believe that the management architecture should allow for application or environment specific management to be integrated with the more generic management functionality it now provides. This could be implemented by facilitating application specific extensions to be plugged into the management system.

We are currently working on extending our management architecture to allow for pluggable management functionalities, possibly using concepts or parts from Marvel or JMX.

Similar mechanisms as we currently use, and the same management architecture is suitable for management of the new generation component models like EJB, COM+ and CORBA Component Model (CCM) [5]. We plan to migrate our current implementation as soon as CCM ORBs become available.

References

1. Mehmet Aksit, Lodewijk Bergmans and Ken Wakita: *An Object-Oriented Model for Extensible Concurrent Systems: The Composition-Filters Approach,* IEEE Transactions on Parallel & Distributed Systems, 1993.
2. Nikolaos Anerousis: *Scalable management services using Java and the World Wide Web,* Ninth Annual IFIP/IEEE International Workshop on Distributed Systems: Operations & Management (DSOM '98), 1998.

3. G.S. Blair, G. Coulson, P. Robin, and M. Papathomas: *An Architecture for Next Generation Middleware*, IFIP International Conference on Distributed Systems Platforms and Open Distributed Processing (Middleware'98), Lake District, UK.
4. Hajo Brunne: *Principle Design Patterns for Manageable Object Request Brokers*, Submission to the OMG CORBA Management Workshop "Vendor/System Integrator Views" Session Monday, 22 September 1997.
5. Cobb, E. et all:*CORBA Components – Volume I*, ed. E. Cobb, Joint revised Submission, OMG TC Document orbos/99-07-01, August 1999.
6. ISO: *ISO 9596: Common Management Information Protocol*. 1991. Geneva.
7. OMG: *The Common Object Request Broker: Architecture and Specification*, formal/98-07-01 (http://www.omg.org).
8. Fabio Costa, Gordon Blair and Geoff Coulson: *Experiments with Reflective Middleware*, ECOOP Workshop on Reflective Object-Oriented Programming and Systems (ROOPS '98), Brussels.
9. ISO: *ISO 10040: Information Technology - Open Systems Interconnection - System Management overview*, 1992.
10. Halldor Fosså and Morris Sloman: *Interactive Configuration Management for Distributed Object Systems*, IEEE Proceedings First International Enterprise Distributed Object Computing Workshop (EDOC '97), Australia, 1997.
11. ISO: *ISO 7498-4: Information processing systems - Open Systems Interconnection - Basic Reference Model - Part 4: Management framework*, 1989.
12. Sun, Java Dynamic Management Kit, http://www.sun.com/software/java-dynamic/.
13. Java Management Extensions, http://java.sun.com/jmx.
14. IONA's Orbix ORB, http://www.orbix.com.
15. Reinhold Kroeger, Markus Debusmann, Christoph Weyer, Erik Brossler, Paul Davern, Aiden McDonald: *Automated CORBA-based Application Management,* DAIS 99, 1999, Helsinki, Finland.
16. Steffen Lipperts, Anthony Sang-Bum Park: *Managing CORBA with Agents*, Interworking 98, 1998, Otawa, Canada.
17. Steffen Lipperts, Dirk Thißen, *CORBA Wrappers for A-posteriori Management: An Approach to Integrating Management with Existing Heterogeneous Systems,* DAIS '99, 1999, Helsinki, Finland.
18. ORBacus ORB, Object Oriented Concepts, http://www.ooc.com
19. Priya Narasimhan, Louise E. Moser, P.M. Melliar-Smith: *Using Interceptors to Enhance CORBA*, IEEE Computer 32[7], p. 62-68, 1999.
20. IONA, OrbixManager, part of OrbixOTM (http://www.iona.com/products/orbixenter/orbixotm/index.html).
21. OMG: Portable Interceptors revised submission, orbos/99-12-02.
22. Günther Rackl: *Multi-Layer Monitoring in Distributed Object-Environments*, second International Working Conference on Distributed Applications and Interoperable Systems (DAIS'99), Helsinki, June 1999.
23. ISO/IEC 10746-3: *Open Distributed Processing – Reference Model (RM ODP), Part 3, Architecture*, 1995.
24. IETF: *RFC 1157 – A Simple Network Management Protocol*. 1990.
25. TeleManagement forum, http://www.tmforum.org/.
26. Maarten Wegdam, Dirk-Jaap Plas, Aart van Halteren, Bart Nieuwenhuis: *ORB instrumentation for the Management of CORBA*, International Conference on Parallel and Distributed Processing Techniques and Applications (PDPTA 2000), June 26-29 2000, Las Vegas, USA.
27. Maarten Wegdam, Aart van Halteren: *Experiences with CORBA interceptors*, position paper for the Workshop on Reflective Middleware, co-located with Middleware 2000, April 2000, New York, USA.

28. Bernd Widmer, Wolfgang Lugmayr: *A comparison of three CORBA Management Tools*, Technical Report, Technical University of Vienna, TUV-1841-99-07.
29. Yasuhiko Yokote: The Apertos Reflective Operating System: The Concept and Its Implementation, OOPSLA'92 Proceedings, October 1992.
30. JacORB is an open source Java ORB developed at the Freie Universität Berlin, http://www.inf.fu-berlin.de/~brose/jacorb/.
31. Distributed Management Task Force, http://www.dmtf.org.
32. Inprise's Visigenic ORB: Visibroker, http://www.visigenic.com/visibroker/.
33. BEA's Weblogic Enterprise, http://edocs.bea.com/wle/index.html.

Evaluation of Constrained Mobility for Programmability in Network Management

Christos Bohoris, Antonio Liotta, George Pavlou

Center for Communication Systems Research
School of Electronic Engineering and Information Technology
University of Surrey, Guildford, Surrey GU2 7XH, UK
{C.Bohoris,A.Liotta,G.Pavlou}@eim.surrey.ac.uk

Abstract. In recent years, a significant amount of research work has addressed the use of code mobility in network management. In this paper, we introduce first three aspects of code mobility and argue that *constrained* mobility offers a natural and easy approach to network management programmability. While mobile agent platforms can support constrained mobility in a rather heavyweight fashion, optimized approaches such as our CodeShell platform presented here can provide performance and scalability comparable to those of static distributed object platforms such as Java-RMI and CORBA. Properly implemented constrained mobility is thus of great importance in network management, resulting in flexible, extensible, programmable systems without prohibitive performance overheads.

Keywords. Code Mobility, Mobile Agents, Java-RMI, CORBA, Performance Evaluation

1 Introduction and Background

Network management has been the subject of intense research over the last decade, with the relevant progress being twofold: on the one hand, approaches and algorithms for solving management problems have been devised; and on the other hand, different management technologies have been proposed and standardized. From the protocol-based approaches of the early 90's, exemplified by the Simple Network Management Protocol (SNMP) [1] and OSI Systems Management (OSI-SM) [2], the focus moved to distributed object-based approaches in the mid to late 90's, exemplified by the Common Object Request Broker Architecture (CORBA) [3] and more recently by Java's Remote Method Invocation (Java-RMI).

The paradigm of moving management logic close to the data it requires is a technique that has been conceived early in the evolution of management architectures, the relevant framework known as "management by delegation" [4]. Subsequent research showed the applicability of this concept in the context of OSI-SM [5] with a similar approach subsequently standardized, the Command Sequencer Systems Management Function (SMF). More recently, the same concept has been proposed in the context of SNMP through the IETF Script MIB [8]. While such approaches are specific to the

A. Ambler, S.B.Calo, and G. Kar (Eds.): DSOM 2000, LNCS 1960, pp. 243 - 257, 2000.

respective management frameworks, the most general approach to delegation in the context of distributed object frameworks is through *object mobility*. Mobile objects are usually termed *Mobile Agents* (MAs) when they act on behalf of other entities and exhibit properties such as autonomy, reactivity, and proactivity.

The emergence of mobile agent frameworks has led many researchers to examine their applicability to network management and control environments. [6] considered code mobility in management and presented a taxonomy of the relevant aspects while [7] discussed the general issues of using mobile agents for network management. Since then a number of other researchers have attempted to use mobile agents in order to solve better specific network management problems. Despite the research efforts until now, there have been little encouraging results regarding the exploitation of "strong" mobility in network management, as defined below.

Mobile agents may move around the network in a reactive or proactive adaptive manner and clone / destroy themselves according to their intelligence. We term this situation "strong mobility" and it is this property that has not yet been shown to achieve better results that static approaches in network management. An alternative possibility for mobile agents is to move from node A to B, typically guided by a "parent" stationary agent, and stay there until their task is accomplished. We term this situation "constrained code mobility" and we believe it is this simpler approach that can be readily exploited in management environments. In this case, instead of predicting the required functionality, standardizing and providing it through static objects in network elements or management systems, mobile code can support it in a dynamic, customizable fashion. The key advantage in this case is that the target node needs only to provide the required "bare-bones" capability which could be dynamically augmented through mobile code, with the additional logic able to change to reflect evolving requirements over time. Such a capability would obviate the use of functionality such as the OSI-SM Systems Management Functions (SMFs) and similar functionality provided in SNMP agents.

The key advantage of constrained code mobility is that it provides the vehicle for programmability through the enhancement of pre-existing functionality in the target node. In [15] we showed how constrained mobility can be used for programmable management systems which can be customized by clients for dynamic connectivity management services. In [9] we showed how the same principle can be used to enhance network elements with add-on functionality for performance monitoring. In both cases we used a general purpose mobile agent platform in order to support constrained mobility. A brief performance comparison with static object platforms in [15] showed that mobile agent solutions are rather heavyweight for constrained code mobility and this led us to the design and implementation of the CodeShell platform (section 3).

In this paper we layout the concepts of constrained, weak and strong code mobility in the context of network management and provide a detailed experimental evaluation of three different approaches to distributed management: 1) static distributed management based on Java-RMI and CORBA respectively as distributed object platforms; 2) dynamic distributed management based on CodeShell, an optimized mobile code platform supporting the constrained mobility paradigm; and 3) dynamic distributed management based on Grasshopper, a general-purpose mobile agent platform.

In section 2 we provide an overview of the general application of code mobility to network management and layout the concepts of constrained, weak and strong mobility. In section 3 we describe the CodeShell platform; this section also serves to identify the architectural aspects and necessary support for constrained code mobility. In section 4 we report on our performance experiments among static distributed object platforms, the CodeShell platform for constrained mobility and the Grasshopper mobile agent platform [14]. We close with our summary and conclusions which show that constrained mobility, if implemented efficiently, leads to performance comparable to the one obtainable with static distributed object solutions, achieving the same level of scalability while providing at the same time an easy and natural approach to programmability with all its associated advantages.

2   Code Mobility in Management

The key benefits code mobility may bring into the network management arena, for each of the five management functional areas, are identified in [7]. These benefits include reduction in network traffic, efficient utilization of computational resources, support for heterogeneous environments, and increased flexibility. Nevertheless, the use of mobile code does not come without costs. In particular, code migration incurs additional traffic into the network, absorbs considerable resources from the agent hosts, and is associated with migration delays of the order of seconds or even tens of seconds, depending on the agent configuration and functionality [16] (see also Section 4.3). Code migration overheads often outweigh its benefits and make this approach inconvenient. It is therefore important to identify the various aspects of code mobility and relate them to network management in order to identify aspects that are particularly beneficial.

In the following subsections we define three different types of code mobility, ranging from the simplest, lightweight form of mobility to the most heavyweight one. For each case we elaborate on its benefits and limitations, identifying advantageous scenarios.

2.1   Constrained Mobility

One of the most elementary forms of code mobility is defined in [6] as Remote Evaluation (REV), after the pioneering work described in [17]. In REV, an application in the client role can dynamically enhance the server capability by sending code to the server. Subsequently, clients can remotely initiate the execution of this code that is allowed to access the resources collocated within the server. Therefore, this approach can be seen as an extension of the client-server paradigm whereby a client in addition to the name of the service requested and the input parameters can also send code implementing new services. Hence the client owns the code needed to perform a service, while the server offers both the computational resources required to execute the service and access to its local resources.

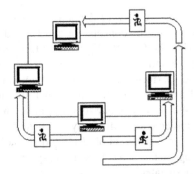

Fig. 1. Constrained Mobility. The agent is created and initialized by a client application and is then shipped to an agent host. The agent execution is then confined to this host.

A natural evolution of the REV model involves sending code not restrained to be a remote service but which can also act as a fully autonomous software entity. This type of code mobility we term *constrained mobility* since the code, upon its creation at a client site, is only allowed to migrate to a remote server where its execution will be confined.

When constrained mobility is adopted in management the code is created by a client acting in the manager role and is, then, dispatched to a target network element acting in an agent/server role (Fig. 1) – in this case the term *agent* is used according to the manager-agent model rather than denoting a mobile agent.

This approach is particularly suited to dynamically programming or upgrading network devices. In this case, the code does not need to be particularly sophisticated. In a simple scenario it could be as simple as collections of objects that can be executed in a remote virtual machine. Therefore, mobility degenerates into a simple dynamic mechanism to efficiently deploy or upgrade network protocols or services.

Code deployment overheads, namely *deployment traffic* and *delay*, represent the drawbacks of general MA approaches. In constrained mobility the code does not need to incorporate complex migration features since its destination is predefined at creation time and is not changed afterwards. As a result, its size and the incurred network traffic are minimal compared to the other forms of mobility (see sections below). Similarly, it will not be necessary to use general purpose MA platforms – usually associated with heavyweight migration mechanisms [16] – and, thus, the code migration time can be considerably reduced (see Section 4.5).

In conclusion, constrained mobility is a particularly well-suited mechanism to dynamically program network elements. It can outperform traditional centralized management for data-intensive tasks and when high degrees of semantic data compression need to be achieved – e.g., through data aggregation or analysis. Constrained mobility is typically advantageous to perform off-line analysis of bulk data and, more generally, to implement tasks whose duration is at least comparable with the overall agent deployment time.

2.2   The Weak and Strong Mobility Models

Weak Mobility

Similarly to constrained mobility, in weak mobility the code is created and initialized by a client application and is, then, shipped to its host. However in the latter, code is not confined to that host since it its meant to perform the same task in more than one location. Mobile agents that follow the weak mobility model do not retain any knowledge of the data processed or of the actions performed in previously visited hosts and, consequently, they can only implement tasks in which this information is not required.

A convenient use of weak mobility is for dynamic decentralization of management tasks that are otherwise performed in a centralized fashion. The agent is delegated part of the management responsibility and will incorporate functionality such as procedures aimed at data semantic compression or aggregation.

A trivial example showing the main advantages of weak mobility is the case in which the management station has to search for a single value in a table, a data structure typically used to store information inside devices. In SNMP management the whole table has to be transferred from the remote element to the management station, where the table rows are searched for the value. Hence, large tables will incur heavy unnecessary traffic into the network and will result in computational overload on the management station.

A more efficient approach is adopted by OSI Systems Management [2], which supports remote scoping and filtering operations. Thus, the searching logic is executed in the device and, consequently, only the retrieved value is transmitted to the manager. The drawback of this approach is that the logic implementing scope and filtering has to be "hard-wired" in the network elements, which tend to become more complex. In addition, such functionality needs to be agreed upon and standardized beforehand.

If constrained mobility were to be used, the agent incorporating the search routine would be shipped to the network device, where it would retrieve the requested values from the local table and return them to the manager. This solution addresses the shortcomings of the OSI approach, retaining its scalability and performance benefits. However, constrained mobility is not suitable in the more general case in which similar tasks are to be run on multiple network elements. In fact, as the number of network elements grows, the management station will be overloaded and the network capacity around it will be saturated by the simultaneous generation and transmission of the agents.

Strong Mobility

With strong mobility, agents are able to access and process data from network elements but can also accumulate information and preserve it upon migration. This feature allows for the implementation of more elaborate tasks in which the agent operations depend on data gathered in previously visited hosts.

In network management, strong mobility is more suited to configuration tasks and to data-intensive tasks involving data aggregation from highly distributed network elements and on-line data analysis. A simple example is a task involving the collection of utilization information from a relatively large number of network elements. In a traditional SNMP-based system, the management station has to poll every single element in order to collect the required raw performance information before it can produce a useful utilization rate. OSI management offers a more efficient mechanism for obtaining the utilization rates, as this is done locally at the network element, but still requires further aggregation of this information at the management station.

Constrained mobility does not suit this task since it would require the deployment of a number of MAs equal to the number of network elements. Each MA would typically be executing for a time negligible with respect to its deployment time and, then, the agent deployment overheads would be unacceptable.

With strong mobility, the agent will be able to preserve the utilization rates of previously visited elements and will then be able to perform a further level of data aggregation independently from the manager. The main drawback associated with strong mobility is the agent size. 'Strong' agents tend to incorporate more intelligence, being larger in size. More critically, the agent size can vary significantly depending on the amount of information that has to be preserved during migration. It is, therefore, important to design the agents in such a way to limit their size variations – e.g., by allowing only semantically compressed information to be carried.

Though in principle there should be uses for weak and strong mobility in network management, research work until now has not resulted in identifying compelling use cases. On the other hand, constrained mobility can be used for network management programmability and we believe that this should be done through optimized platforms such as our CodeShell one presented in the next section.

3   The CodeShell Prototype Platform for Constrained Mobility

3.1   Introduction

Constrained mobility requires at least the following two important facilities:

- A mechanism for migrating management logic along with initial parameters to a destination machine hosting the necessary resources.
- A naming service in order to distinguish between objects and also bind one object to another.

Mobile agent platforms include both these features but also many other features that allow them to be used as general purpose solutions for weak and strong mobility. As such, they are rather heavyweight and have scalability problems. On the other hand, the performance of distributed object frameworks such as Java-RMI and CORBA is more acceptable but these support only static objects that can communicate remotely with other objects in different network nodes. In order to build constrained mobility applications, a distributed object framework's communication and naming service facilities could be re-used. In addition, a thin layer of functionality is needed, allowing the migration of management logic and supporting naming and binding of objects. As a means of validating this approach, the CodeShell prototype platform was designed and implemented providing such facilities over Java-RMI. The CodeShell platform supports constrained mobility and it is specifically tailored to network management.

Constrained mobility is particularly efficient for providing a number of programmable, customizable network management services. In the case of performance management for example, in order to collect performance data intended for off-line analysis, a performance monitor can be conveniently transferred near the resources of the node that needs to be monitored. The transfer of logic in this case allows the easy customization of the monitoring activities and the modification of monitoring characteristics in a dynamic way. In fault management, constrained mobility can be used to send event filtering and correlation logic to the node that is monitored for faults. Performance and fault management services along with a configuration management system, can be integrated and provided to customers in a dynamic, customizable fashion using the constrained mobility model as described in [15].

3.2 Design and Implementation

The CodeShell platform consists of the following components:

- CodeShell Communication Service (CCS): Allows the communication between remote objects. It is also responsible for the transfer of byte-code between two remote machines.
- CodeShell Naming Service (CNS): Provides functionality related to the names assigned to objects. All names and important object information are stored in a local database. This allows the lookup of objects and the binding between them for local communication.
- CodeShell Core System: Coordinates the operation of an individual CodeShell. Also provides a single interface allowing CodeShell objects to perform CCS and CNS related operations.
- Base CodeShell Object: Provides basic functionality that should be inherited by any object intended for use inside a CodeShell. Most functionality relates to basic CCS and CNS related operations.
- CodeShell Textual User Interface: A user interface that allows the user to manage (create, delete, list, etc.) objects within a CodeShell. This component is completely pluggable and can be easily detached from the main CodeShell system and replaced with an alternative environment.

A minimum typical scenario of operation involves two remote machines. One Java-RMI registry and one CodeShell are initialized on each machine (see Fig. 2). In the CodeShell of the machine in the client role, a "master" CodeShell object is created. This object will then contact the CodeShell's CCS in order to send byte-code containing management logic to the remote machine {Step 1}. The CCS will then locate the CCS component of the remote CodeShell and transfer the byte-code and a list of initial parameters {Step 2}. When the byte-code arrives at the machine in the server role, a "logic" object is created from it and it is initialized with the provided parameters. In this CodeShell a "target" object is already created by the user waiting to provide the necessary raw resources. The "logic" object contacts the local CNS and performs a lookup for the "target" object {Step 3}. When this is located, the two objects are bound so that they can communicate locally with each other {Step 4}. The logic object is now ready to perform its task using the target object to obtain the necessary raw information {Step 5}. When it is time for a report to be sent back to its "master" object it will contact the CCS {Step 6} which will transfer it to the remote CodeShell {Step 7}. From there the local CCS is responsible for the report to be passed to the "master" object {Step 8}. The CodeShell platform was developed in Java using Sun's JDK and uses the runtime environment version 1.2.1.

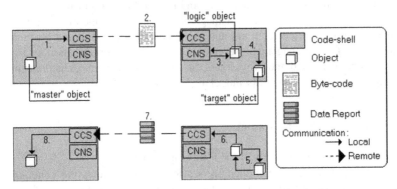

Fig. 2. A typical constrained mobility application scenario running inside two remote Code-Shells.

4   Experimental Evaluation

4.1   Case Study

In order to evaluate the performance overheads of Java-RMI, CORBA, CodeShell and Grasshopper respectively, we used the performance monitoring case study described in detail in [9]. The aim here is to provide traffic rates with thresholds, quality of service alarms and periodic summarization reports by simply observing raw information such as traffic counters in network elements. This is functionality similar to the OSI-SM metric monitoring and summarization facilities (X.739/X.738) [10][11] but when it is provided through code mobility, users of the service are also able to customize it according to the semantics of a particular application as explained in [9].

When constrained code mobility is deployed using either CodeShell or Grasshopper, a performance monitor object is created by a "master" object somewhere and is sent to execute within a target node. When Java-RMI and CORBA are used, the performance monitor object is created at the target node through an object factory and the relevant intelligence needs to pre-exist at that node. The performance monitor gathers information locally, applies thresholds and sends QoS alarms or periodic summarization reports to the master object.

The "target" object is actually an adapter for the underlying SNMP. For its implementation and operation the AdventNet SNMP version 2.0 libraries were used in order to query an SNMP agent for raw performance information. Given this functionality, we are interested to measure the creation/migration overheads, the cost of remote invocation that models the reporting of results, both in terms of response time and packet sizes, and the computing requirements at a the target node. Though our case study is specific to a particular problem domain i.e. performance management, the described measurements are general enough to give us insight on the overheads of the constrained mobility approach in network management in general.

4.2 Method

The performance monitoring system has been implemented over four different infrastructures, the Grasshopper mobile agent platform, Java-RMI, CORBA, and Code-Shell. The aim was to assess the impact that these underlying technologies may have on the monitoring system.

Grasshopper is not one of the most efficient MA platforms. It has been chosen because, due to its functionality, it can be considered a general-purpose MA platform. Moreover, Grasshopper follows the current standardization directions, since it is compliant with both MASIF [13] and FIPA [12].

Java-RMI and CORBA have been chosen as representative of the most popular 'static' distributed object technologies. In addition, CORBA is emerging as a significant technology for network and service management.

Finally, CodeShell is our implementation of a platform supporting constrained code mobility. It is an approach that lays in between two extreme solutions. In the first case the monitoring system is enabled with strong code mobility, which is supported by a general purpose MA platform. In the second case, the systems cannot rely on any form of code mobility since it is entirely based upon static distributed object technologies. For this reason, CodeShell can be regarded as an optimized version of a general-purpose MA platform, which aims at achieving performance comparable to the one obtained with static distributed object technologies.

During our experiments we have run the same performance monitoring tasks over the four different platforms, measuring in each case the total response time, the traffic incurred in the network and the total memory requirements. In the cases of Grasshopper and CodeShell the measurements have been taken at steady state, that is after the code had been shipped to the remote elements. In this way we could perform a direct comparison with the implementations based on Java-RMI and CORBA, respectively. The additional overheads incurred by code mobility – namely code deployment time and network traffic incurred during the transmission of the code from manager to network elements – were measured separately.

In order to measure the total remote invocation response times, timestamps were taken using the *currentTimeMillis* method of the *java.lang.System* class. An array of objects (class *java.util.Vector*) containing 25 numbers of type *java.lang.Double* was remotely transferred 100 times between two objects located in different machines, measuring the total transmission time. The same procedure was repeated while increasing the number of elements in the list to 50, 75, and 100. This operation in fact models the periodic summarization reports generated and remotely sent by the entity equipped with the performance monitoring logic.

For the same experimental cases and in order to calculate the total incurred traffic, we measured the TCP packet sizes using the *tcpdump* program originated at the Lawrence Berkeley laboratory, reporting the total payloads at the TCP level.

To measure the memory requirements, program sizes were measured for the server side of the Grasshopper, CodeShell and Java-RMI systems. The measurements were taken using the *totalMemory*, and *freeMemory* methods of the *java.lang.Runtime* class. The first method provides the total amount of memory allocated by the Java Virtual Machine (JVM). The second one returns an approximate value of the amount of memory left free inside the JVM – i.e., memory available for future object allocation. The difference of these two values provides the amount of memory required by the performance monitoring system under evaluation which includes the required platform related classes in addition to the "target" and to the "logic" objects.

The experiments have been repeated 100 times for each of the above cases in order to perform a statistical analysis and study the significance of the measurements. The experiments where carried out using two different machines over a lightly loaded 100 Mbit/sec Ethernet in the role of the management network with the following specification: Sun Microsystems Ultra-10, 256MB of memory, Sun's Solaris 2.5.1 version of UNIX.

4.3   Response Times Measurements

The response time of management operations for each of the four cases of performance monitoring systems are reported in Fig. 3 which, for an easier comparison, combines in a single chart the mean values and best liner fit of the results.

The first conclusion that can be drawn by observing the plots is that the system based on CodeShell exhibits the same degree of scalability as the one of the systems based on Java-RMI and CORBA. In fact, the slopes of the curves of those three cases have a comparable value. On the contrary, the Grasshopper system exhibited a much bigger slope showing its intrinsic inability to perform well under more demanding conditions.

From the performance point of view the CodeShell system gave a response time in the order of 2-3 times larger than the one of the Java-RMI system and in the order of 4 times larger than the one of the CORBA system.

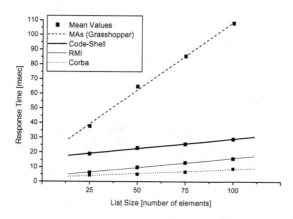

Fig. 3. Mean values and best linear fit of response times.

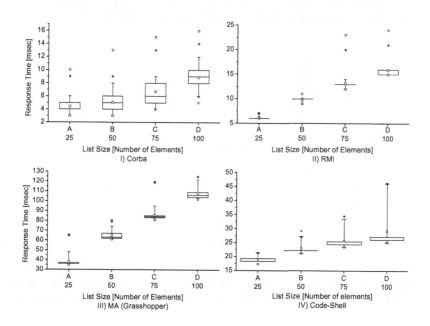

Fig. 4. Statistical Box Charts showing response times for each of the four experimented cases. The boxes include the 25-75% boundaries, the mean values (a black square) and the median values (a line). The 5-95% range boundaries are delimited by whiskers. The outliers are depicted with circles and stars.

The conclusions based on Fig. 3 have been validated by statistical analysis. Fig. 4 depicts the results of this analysis in the form of statistical box charts. It should be noted that these boxes do not have overlapping values and this fact leads us to the conclusion that the response times of the four different solutions are indeed statistically different. Therefore, the mean values and the line slopes of Fig. 3 are statistically representative and can be employed to carry out the above comparative analysis.

4.4   Traffic Measurements

We also measured the packet sizes in all four cases. An array of objects (class *java.util.Vector*) containing 25, 50, 75 and 100 *"Double"* numbers respectively was remotely sent using remote invocations in the Mobile Agent, CodeShell, RMI and CORBA systems. Each time, the payload of the TCP packets was measured. A chart of the results gathered can be seen in Fig. 5.

It is interesting to observe that, within the measured range of values, the four solutions incurred a comparable level of traffic. The Grasshopper and RMI systems performed better for small scales whilst the CodeShell and CORBA systems exhibited better performance for larger scales. The CodeShell platform transparently optimizes the transfer procedure and this is the reason why, for a large number of elements, it incurred less traffic in the network compared with the standard RMI system.

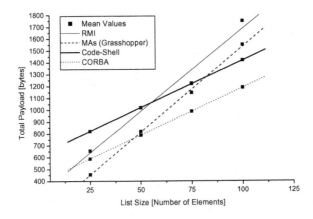

Fig. 5. Mean and best linear fit of total incurred TCP payloads, measured as the sum of all the bytes incurred in the network to complete the given network performance monitoring task.

4.5   Memory Measurements

The memory requirements for the monitoring systems based on Grasshopper, Java-RMI, and CodeShell, are compared in Fig. 6.

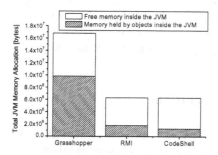

Fig. 6. Memory requirements for the Java-based network performance monitoring systems.

It can be observed that CodeShell performs as well as Java-RMI and significantly better than Grasshopper. The latter, resulted in a fivefold occupation of memory which is yet another dramatic drawback of relying on a general-purpose MA platform.

4.6 Code Migration Overheads

The delay and traffic involved during code migration have been measured for the two programmable network performance monitoring systems based on Grasshopper and CodeShell, respectively. The times involved in the migration of performance monitoring logic from the manger to the network elements are reported in Fig. 7 in the form of statistical box charts.

By substituting the generic code migration mechanism of Grasshopper with a simpler code deployment protocol we have been able to reduce by four times the time required to program a network element. The fact the statistical boxes do not overlap proves that the difference in code migration time between the two approaches is statistically significant.

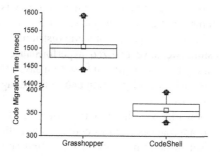

Fig. 7. Statistical Box Charts showing code migration times involved in the programmable network performance monitoring systems. The boxes include the 25-75% boundaries, the mean values (a black square) and the median values (a line). The 5-95% range boundaries are delimited by whiskers. The outliers are depicted with circles and stars.

We also measured the additional traffic incurred by code migration. The transmitted data for the CodeShell system was 2,236 bytes; for the Grasshopper system it was 2,854 bytes. There was no need to repeat the measurements since we were able to measure the exact payload by discriminating it from the background traffic.

5   Discussion and Conclusions

Over the last three years we have been engaged in the task of designing and implementing an integrated Network Management system based on the use of Mobile Agent technologies. Our investigation led us to conclude that general-purpose MA platforms are not a viable infrastructure over which dynamic, programmable management systems can be realized. This is due to the fact that MA platforms tend to be rather heavyweight and do not scale well. In particular, code migration involves delays which in several cases are orders of magnitude larger than the timescales typical of network management systems. Moreover, following the MbD idea originated in 1991, nearly ten years of discussions in the management community have failed to identify a single case in which the use of strong mobility can have a significant impact on NM applications.

On the contrary, we believe that constrained mobility can be the vehicle to realize network programmability and MbD functionality with current technologies.

These observations have induced us to shift our investigation towards the constrained mobility concept in order to assess more precisely its effectiveness in the specific context of network management. We have implemented CodeShell in order to establish whether it was possible to realize management systems based on constrained mobility, achieving at the same time performance levels comparable to the ones of systems based on the most popular static distributed object technologies. We also aimed at quantifying the performance gain achievable by giving up on general-purpose MA technologies, retaining only the most basic form of code mobility exemplified by constrained mobility.

The results presented herein suggest that constrained mobility is easily integrated in network management systems. Moreover, using constrained mobility it is still possible to achieve performance and scalability typical of static distributed object technologies.

To draw these conclusions we have realized a performance monitoring systems over CodeShell, a constrained mobility platform based on Java-RMI. We believe that other management functions such as configuration and fault management can similarly benefit from constrained mobility.

While mobile agent frameworks were initially thought as rivals to static distributed object frameworks, the two approaches need to coexist. We believe that constrained mobility is the required level of code mobility that can find concrete application in network management. Real synergy could be achieved if stationary agents could be provided using static objects, with method invocations being possible between mobile and static objects in both directions. Such an environment would combine the best of both worlds but it is not clear at present whether this type of seamless integration is achievable.

Acknowledgments. This work was undertaken in the context of the IST MANTRIP (MANagement Testing and Reconfiguration of IP based networks using Mobile Software Agents) project, which is partially funded by the Commission of the European Union.

References

1. J.Case, M.Fedor, M.Schoffstall, J.Davin, *A Simple Network Management Protocol (SNMP)*, IETF RFC 1157, 1990.
2. ITU-T Rec. X.701, Information Technology - Open Systems Interconnection, *Systems Management Overview*, 1992.
3. Object Management Group, *The Common Object Request Broker: Architecture and Specification (CORBA)*, Version 2.0, 1995.
4. Y. Yemini, G. Goldszmidt, S. Yemini, *Network Management by Delegation*, in Integrated Network Management II, Krishnan, Zimmer, eds., pp. 95-107, Elsevier, 1991.
5. N. Vassila, G. Pavlou, G. Knight, *Active Objects in TMN*, in Integrated Network Management V, Lazar, Saracco, Stadler, eds., pp. 139-150, Chapman & Hall, 1997.
6. M. Baldi, S. Gai, G.P. Picco, *Exploiting Code Mobility in Decentralised and Flexible Network Management*, Proc. of the 1st International Workshop on Mobile Agents 97 (MA'97), Berlin (Germany), K. Rothermel and R. Popescu-Zeletin eds., April 1997, Springer, Lecture Notes on Computer Science vol. 1219, ISBN 3-540-62803-7, pp. 13-26.
7. A. Bieszczad, B. Pagurek, T. White, *Mobile Agents for Network Management*, IEEE Communications Surveys, Vol. 1, No. 1, http://www.comsoc.org/pubs/ surveys, 4Q1998.
8. J. Schoenwaelder, J. Quittek, *Script MIB Extensibility Protocol Version 1.0*, RFC 2593, May 1999.
9. C. Bohoris, G. Pavlou, H. Cruickshank, Using Mobile Agents for Network Performance Management, Proc. of the IEEE/IFIP Network Operations and Management Symposium (NOMS '00), Hawaii, USA, J. Hong, R. Weihmayer, eds., pp. 637-652, IEEE, April 2000.
10. ITU-T Rec. X.739/X.738, Information Technology - Open Systems Interconnection, *Systems Management Functions - Metric Objects and Attributes/Summarization Function, 1992/1993.*
11. G. Pavlou, G. Mykoniatis, J. Sanchez, *Distributed Intelligent Monitoring and Reporting Facilities*, IEE Distributed Systems Engineering Journal (DSEJ), Special Issue on Management, Vol. 3, No. 2, pp. 124-135, IOP Publishing, 1996.
12. Foundation for Intelligent Physical Agents, web page: http://www.fipa.org/ .
13. Object Management Group, *Mobile Agent System Interoperability Facilities Specification*, orbos/97-10-05, 1997, ftp://ftp.omg.org/pub/docs/orbos/97-10-05.pdf
14. The Grasshopper Agent Platform http://www.ikv.de/products/grasshopper/index.html.
15. D. Griffin, G. Pavlou, P. Georgatsos, *Providing Customisable Network Management Services Through Mobile Agents*, Proc. of the 7th International Conference on Intelligence in Services and Networks (IS&N'00), Athens, Greece, G. Stamoulis, A. Mullery, D. Prevedourou, K. Start, eds., pp. 209-226, Springer, February 2000.
16. G. Knight, R. Hazemi, *Mobile Agents based Management in the INSERT Project*, Journal of Network and Systems Management, Vol.7, No.3, pp. 271-293, September 1999.
17. J.W. Stamos, D.K. Gifford, *Remote Evaluation*, ACM Transactions on Programming Languages and Systems. 12(4), pp.537-565, October 1990.

Author Index

Lecture Notes in Computer Science

For information about Vols. 1–1887
please contact your bookseller or Springer-Verlag

Vol. 1921: S.W. Liddle, H.C. Mayr, B. Thalheim (Eds.), Conceptual Modeling for E-Business and the Web. Proceedings, 2000. X, 179 pages. 2000.

Vol. 1922: J. Crowcroft, J. Roberts, M.I. Smirnov (Eds.), Quality of Future Internet Services. Proceedings, 2000. XI, 368 pages. 2000.

Vol. 1923: J. Borbinha, T. Baker (Eds.), Research and Advanced Technology for Digital Libraries. Proceedings, 2000. XVII, 513 pages. 2000.

Vol. 1924: W. Taha (Ed.), Semantics, Applications, and Implementation of Program Generation. Proceedings, 2000. VIII, 231 pages. 2000.

Vol. 1925: J. Cussens, S. Džeroski (Eds.), Learning Language in Logic. X, 301 pages 2000. (Subseries LNAI).

Vol. 1926: M. Joseph (Ed.), Formal Techniques in Real-Time and Fault-Tolerant Systems. Proceedings, 2000. X, 305 pages. 2000.

Vol. 1927: P. Thomas, H.W. Gellersen, (Eds.), Handheld and Ubiquitous Computing. Proceedings, 2000. X, 249 pages. 2000.

Vol. 1928: U. Brandes, D. Wagner (Eds.), Graph-Theoretic Concepts in Computer Science. Proceedings, 2000. X, 315 pages. 2000.

Vol. 1929: R. Laurini (Ed.), Advances in Visual Information Systems. Proceedings, 2000. XII, 542 pages. 2000.

Vol. 1931: E. Horlait (Ed.), Mobile Agents for Telecommunication Applications. Proceedings, 2000. IX, 271 pages. 2000.

Vol. 1658: J. Baumann, Mobile Agents: Control Algorithms. XIX, 161 pages. 2000.

Vol. 1766: M. Jazayeri, R.G.K. Loos, D.R. Musser (Eds.), Generic Programming. Proceedings, 1998. X, 269 pages. 2000.

Vol. 1791: D. Fensel, Problem-Solving Methods. XII, 153 pages. 2000. (Subseries LNAI).

Vol. 1799: K. Czarnecki, U.W. Eisenecker, Generative and Component-Based Software Engineering. Proceedings, 1999. VIII, 225 pages. 2000.

Vol. 1812: J. Wyatt, J. Demiris (Eds.), Advances in Robot Learning. Proceedings, 1999. VII, 165 pages. 2000. (Subseries LNAI).

Vol. 1932: Z.W. Raś, S. Ohsuga (Eds.), Foundations of Intelligent Systems. Proceedings, 2000. XII, 646 pages. (Subseries LNAI).

Vol. 1933: R.W. Brause, E. Hanisch (Eds.), Medical Data Analysis. Proceedings, 2000. XI, 316 pages. 2000.

Vol. 1934: J.S. White (Ed.), Envisioning Machine Translation in the Information Future. Proceedings, 2000. XV, 254 pages. 2000. (Subseries LNAI).

Vol. 1935: S.L. Delp, A.M. DiGioia, B. Jaramaz (Eds.), Medical Image Computing and Computer-Assisted Intervention – MICCAI 2000. Proceedings, 2000. XXV, 1250 pages. 2000.

Vol. 1937: R. Dieng, O. Corby (Eds.), Knowledge Engineering and Knowledge Management. Proceedings, 2000. XIII, 457 pages. 2000. (Subseries LNAI).

Vol. 1938: S. Rao, K.I. Sletta (Eds.), Next Generation Networks. Proceedings, 2000. XI, 392 pages. 2000.

Vol. 1939: A. Evans, S. Kent, B. Selic (Eds.), «UML» – The Unified Modeling Language. Proceedings, 2000. XIV, 572 pages. 2000.

Vol. 1940: M. Valero, K. Joe, M. Kitsuregawa, H. Tanaka (Eds.), High Performance Computing. Proceedings, 2000. XV, 595 pages. 2000.

Vol. 1941: A.K. Chhabra, D. Dori (Eds.), Graphics Recognition. Proceedings, 1999. XI, 346 pages. 2000.

Vol. 1942: H. Yasuda (Ed.), Active Networks. Proceedings, 2000. XI, 424 pages. 2000.

Vol. 1943: F. Koornneef, M. van der Meulen (Eds.), Computer Safety, Reliability and Security. Proceedings, 2000. X, 432 pages. 2000.

Vol. 1945: W. Grieskamp, T. Santen, B. Stoddart (Eds.), Integrated Formal Methods. Proceedings, 2000. X, 441 pages. 2000.

Vol. 1948: T. Tan, Y. Shi, W. Gao (Eds.), Advances in Multimodal Interfaces – ICMI 2000. Proceedings, 2000. XVI, 678 pages. 2000.

Vol. 1952: M.C. Monard, J. Simão Sichman (Eds.), Advances in Artificial Intelligence. Proceedings, 2000. XV, 498 pages. 2000. (Subseries LNAI).

Vol. 1954: W.A. Hunt, Jr., S.D. Johnson (Eds.), Formal Methods in Computer-Aided Design. Proceedings, 2000. XI, 539 pages. 2000.

Vol. 1955: M. Parigot, A. Voronkov (Eds.), Logic for Programming and Automated Reasoning. Proceedings, 2000. XIII, 487 pages. 2000. (Subseries LNAI).

Vol. 1960: A. Ambler, S.B. Calo, G. Kar (Eds.), Services Management in Intelligent Networks. Proceedings, 2000. X, 259 pages. 2000.

Vol. 1961: J. He, M. Sato (Eds.), Advances in Computing Science – ASIAN 2000. Proceedings, 2000. X, 267 pages. 2000.

Vol. 1963: V. Hlaváč, K.G. Jeffery, J. Wiedermann (Eds.), SOFSEM 2000: Theory and Practice of Informatics. Proceedings, 2000. XI, 460 pages. 2000.

Vol. 1966: S. Bhalla (Ed.), Databases in Networked Information Systems. Proceedings, 2000. VIII, 247 pages. 2000.

Vol. 1967: S. Arikawa, S. Morishita (Eds.), Discovery Science. Proceedings, 2000. XII, 332 pages. 2000. (Subseries LNAI).

Vol. 1968: H. Arimura, S. Jain, A. Sharma (Eds.), Algorithmic Learning Theory. Proceedings, 2000. XI, 335 pages. 2000. (Subseries LNAI).

Vol. 1969: D.T. Lee, S.-H. Teng (Eds.), Algorithms and Computation. Proceedings, 2000. XIV, 578 pages. 2000.

Vol. 1970: M. Valero, V.K. Prasanna, S. Vajapeyam (Eds.), High Performance Computing – HiPC 2000. Proceedings, 2000. XVIII, 568 pages. 2000.

Vol. 1971: R. Buyya, M. Baker (Eds.), Grid Computing – GRID 2000. Proceedings, 2000. XIV, 229 pages. 2000.

Vol. 1975: J. Pieprzyk, E. Okamoto, J. Seberry (Eds.), Information Security. Proceedings, 2000. X, 323 pages. 2000.

Vol. 1976: T. Okamoto (Ed.), Advances in Cryptology – ASIACRYPT 2000. Proceedings, 2000. XII, 630 pages. 2000.